Walking Through Fire

*An Iwo Jima Survivor's
Remembrance*

Walking Through Fire

An Iwo Jima Survivor's

5/28/09

Remembrance

Arvy A. Geurin

Arvy Albin Geurin
(US Navy – 1943-1945)

As told to Gale W. Geurin

McKenna Publishing Group

San Luis Obispo, California

Walking Through Fire

ISBN: 978-1-932172-31-7
LCCN: 2008940600

Cover design by Leslie A. Parker

First Edition
10 9 8 7 6 5 4 3 2 1
Printed in the United States of America

Visit us on the Web at: www.mckennapubgrp.com

Dedicated To

*The brave men of the USS Napa
and the Fourth Marines
who fought and died on Iwo Jima*

In Memory Of

Lt. (jg) Donald Ernest Ritche, USNR, Boat Group Officer, USS NAPA
19 February 1945 - Killed in action during the invasion of Iwo Jima. He was in command of, and led, the eighth wave to a successful landing on Blue Beach when hit by a burst of machine gun fire inflicting wounds which later caused his death. He was buried at sea on 21 February with full military honors from the USS Newberry.

Seaman 2nd Class Harold Warren Hornick, USNR, USS NAPA
19 February 1945 - A member of the USS NAPA Beach Party, he was killed in action on the beach at Iwo Jima. He was hit by enemy gun fire while carrying out his orders. He is buried on the small island that cost so much in human life.

Fireman 1st Class Anthony Alfonso Morrone, USNR, USS NAPA
20 February 1945 - An engineer member of boat crew LCM26, he was killed in action during the invasion of Iwo Jima. He died of wounds received from machine gun fire while his boat was landing on the beach. He was buried at sea with full military honors.

Seaman 1st Class James Carlton Owens, USNR, USS NAPA
19 February 1945 - Member of boar crew LCM26, killed in action during the invasion of Iwo Jima. He was reported missing in action after being wounded by machine gun fire. His death was later verified by the Bureau of Naval Personnel.

Seaman 1st Class John Max Reed, USNR, USS NAPA
19 February 1945 - A member of the USS NAPA Beach Party, he was killed in action on the beach at Iwo Jima. He was hit by mortar fire on the second terrace of the beach while acting as a stretcher bearer. He is buried on Iwo Jima.

Seaman 1st Class Benjamin Charlie Schlabach, USNR, USS NAPA
7 August 1945 - He was killed while manning his battle station during an enemy air raid at Okinawa. A military funeral was held on board the USS NAPA by the ship's Chaplain. He was buried in Okinawa.

Lt (jg) Ford Eshleman, MC, USNR, USS NAPA
Missing in Action. He served on the USS NAPA from November 1944 through the Iwo Jima engagement. He was later transferred to destroyer duty and was serving on board the destroyer USS BRAINE when she was severely damaged by three kamikaze planes.

IWO JIMA

They went down to the sea,
brave men, all,
to the black sands,
and cluttered beaches,
to slippery terraces,
and rabbit warrens,
to reach the mountain,
to raise the flag,
to freedom,
brave men, all.

Dedicated to the 4th Marines and Men of the USS NAPA

Brave Men All

Gale W. Geurin

Fighters were strafing the dark island and relentlessly bombing its highest point, Mt. Suribachi. In the amtrack landing craft, I could see much of the bombing and shelling of the island, and I could hear the terrible sound of guns, from the 16-inch guns down to the five-inch guns. The burst of the 40-mm guns echoed— **Boom! Boom! Boom!** *It was a demonic rhythm filling my ears. The ships behind me fired over our heads, exploding shells on the island before us. The firing of the mortars split the air above me. Waves crashed into the craft, hitting us in the face as we stared ahead at the dark island. Thick, black flack scarred the sky. The island had been bombed for six months before the invasion, and shelled ceaselessly with the ships' big guns. We were expecting a speedy mop-up. Our Lieutenant had bragged how this would be a quick, souvenir-gathering operation, and off to Guam. There could not be any resistance left.*

Now, in the LST (landing ship, tank) with more bombs zeroing in on Iwo Jima and the ships' shells pounding the beach, I wasn't thinking of a mop-up operation. I was eager to get on that island. Anxious to be a part of this war. This . . . this is why I had joined! In spite of all this, it didn't dawn on me that I could actually be killed on that hunk of volcanic rock in the Pacific. At nineteen, even in the midst of war, I was invincible. As the amtrack bounced on the choppy sea, and the big guns **boomed** *overhead, I wasn't thinking of the battle ahead, but how in two short years, life had changed from planning a summer of fun after graduation to being here, splattered by bullet-whipped waves, heading toward that small, seemingly insignificant island in the Pacific.*

CHAPTER 1
Bakersfield

"Hey, Arv, check that fryer." It was my Uncle Vernon Edison Smith's voice, still as loud and clear in my mind as it was back in April 1941 in Bakersfield, California. That tiny far-off island in the Pacific wasn't even a dot on a map back then to me when the only worry I had was whether or not the doughnuts in my uncle's bakery shop were made right. I was the doughnut maker and fryer that summer before my junior year at East Bakersfield High School. My goal in life back then was to become as good a baker as my uncle. I loved the smell and taste of doughnuts coming out of the fryer.

At the high school, I ploughed through the classes, not very interested in the academics, doing my best because it was both expected and required of me to pursue my passion. My real interest lay in athletics, especially swimming. It was there that I won three medals in competition, including freestyle, and felt great when my classmates and school cheered me on. Later, in the Navy, my swimming skills would be literally put to the test.

"Got it! Coming out!" I shouted back at Uncle Vernon. I gripped the basket laden with fresh doughnuts with both hands and heaved it effortlessly onto the spillout tray. The plain, cake doughnuts tumbled down, and settled, waiting for the final touches . . . icing and decorations.

Steaming hot donuts tumbled against one another, with each bounce sending another round of sweet, tantalizing aromas filling the air. I grabbed one of the large pots of chocolate icing and a large metal tongs and snatched a donut, dunked it, slapped it on wax paper, and kept at it until all the donuts on the spillout tray had been iced.

My cousin V.E., two years older than I, who was Uncle Vernon's son and namesake, picked up the big, tin shaker of sprinkles and shook it over the freshly iced donuts so they were liberally speckled with colorful candy sprinkles.

One day, V.E. and I would open our own bakery. One day. Far in the future. For now, we worked for, and learned from Uncle Vernon. That was my focus back then. How soon . . . too soon . . . all of that would change one bright Sunday morning in December.

The amtrack rose on the incoming waves and crashed hard upon the sea as each wave rolled passed. We were packed in tighter than sardines in a can, and the whiplash of the rolling ocean barely moved us. It was impossible to think about anything but the shells whizzing overhead or what waited on the shore. The closer we got to the island, the more the reality of the war closed in on me. It wasn't so long ago that I was still in Bakersfield, with the only concern was how to become the perfect baker. How was it an apprentice baker from California ended up on a crowded, bouncing boat heading toward the hellfire of war? I looked around me. We all were ordinary young men . . . not so far removed from being just teenage boys who should be looking forward to summer fun and summer jobs and summer romances. We weren't action heroes or extraordinary men. We were farm boys and city jocks; scholars and drop-outs; rich and poor; we were just young boys brought together by a common goal. Just a year ago, how could any of us had imagined going off to war like this? Yet, here we were altogether in an amtrack moving toward the hungry jaws of war, to what would become the bloodiest battle of the Pacific.

In my uncle's bakery, it wasn't just doughnuts we made that filled the air with their enticing aromas and tempting tastes. There were also the fresh baked breads. When those loaves were done, I liked taking a hot loaf of bread from the bakery oven, scooping out the soft white dough, smearing the crust with sweet butter, and scarffing it down while it was still hot. My cousin V.E. and I delighted in doing that. It was a small thing that I never once thought twice about, until it wasn't possible anymore. How short a time later it was that the bakery was but a poignant memory, and V.E. was gone forever.

Uncle Vernon, V.E.'s father, deliberately looked the other way when V.E. and I snatched a loaf of hot bread. V.E. winked, and tossed the hollowed out loaf at me. I caught it with one hand and grabbed the deep tub of butter with the other. As I slathered the crusty hot bread with a blanket of butter, V.E. talked about what he would be doing after he graduated that year. "I've been thinking about taking a year off from the bakery after graduation," V.E. said as he watched the golden butter instantly melt on the bread. "Travel, maybe." Going to war was the farthest thing from either of our minds.

"Oh, yeah?" I snickered. "Travel where? How?" Neither V.E. nor I had the funds to go world traveling. I broke the crusty bread in half and threw half of it his way.

He deftly caught it. "I dunno," V.E. replied, "maybe hitchhike around the country. Be fun, you know?" He shrugged. "Safe and cheap way to go." We did not even imagine how far, and how dangerously, V.E. would travel, less than a year later.

Jimmy, V.E.'s younger brother, a year younger than I, came in just as I bit into my half of the bread. "At it again, huh?" Jimmy remarked and laughed. V.E. playfully slapped him across the head, and held his half of the buttery, hot bread high over his head. Jimmy jumped for it, but missed. V.E. was tall, lanky, full of life and laughter.

Just then, Uncle Vernon yelled, "Jimmy, get your butt in here!" V.E. and I grinned at each other and finished our purloined snack, as Jimmy hopped-to. Then back to work at the ovens.

Back then, in Bakersfield, California, I was already a long ways from "home," and couldn't imagine being further, to a little sulfur island far in the Pacific. I doubt if at the time of my birth in December of 1925 in Hot Springs, Arkansas, my parents could have ever foreseen either of their sons ever going off to war. Dad had been drafted into the Army into the *war to end all wars*. He was confident the world would never again experience such terrible destruction. World War I did not have the Roman numeral designation then; it was supposed to be remembered as the only worldwide war, and involved over 150 countries on four continents and numerous islands in the Pacific, Atlantic and Indian Oceans. It was supposed to settle borders and leaderships and territorial disputes.

In the uneasy peace that followed, world leaders were determined not to ever have it happen again.

CHAPTER 2
Journeys

There in that crowded landing craft heading toward the black sands of Iwo Jima, the bakery and the comfortable banter between my cousins faded, and I thought of my father as a young man. I had voluntarily joined the Navy to keep from being drafted into the Army. My father hadn't had that choice. In 1918, he had been drafted into the service, and was sure he would be sent into the fray. He had been newly married at the time. There was fierce fighting overseas, and the casualties were high on both sides. He hadn't joined willingly, but neither did he resist. He was ready for whatever might come, but it was only by a stroke of fate that he had stayed stateside. He became an Army guard over German prisoners of war held in the States. I wondered if he felt back then as I did as the amtrack moved toward shore, sure of himself, not yet afraid of the journey that lay ahead, waiting for that call that would put him in the middle of battle.

Journeys ... life is full of them, each one having crossroads that change the course of lives depending on the road chosen. The journey to Bakersfield and the bakery began in Arkansas in 1918 when a 23-year-old young farmer named Hilbert Haywood Geurin won the hand of, and married, a merchant's young daughter called Willie Octavia Smith, who was but 19 at the time. It was a journey of thousands of miles and idealistic dreams away from California, and an impossible journey a world away from a war that had not even started, or an island not even known. Three years later, their first son, my brother Elton Edison was born; four years later, I came along. During those seven years, the farm boy and the merchant's daughter worked hard together, and journeyed a long ways from their beginnings in Arkansas.

My father was a product of the times when education was considered a luxury, not a necessity. He managed to finish eighth grade in a one-roomed school house in the back country of Arkansas, an outstanding accomplishment for a backwoods Arkansas boy. In spite of that, he provided well for his family. In those times what counted was a willingness and ability to work. My father had both of those in spades.

My mother only acquired a third-grade education. Her father pulled her out of school to help out in the general store he owned in Hot Springs, Arkansas. An education was not considered necessary for women during that time. The only thoroughly educated women in that part of the country were the schoolmarms. Store clerks didn't need an education, in the view of my grandfather and most people of the time. They only needed to know the product. That my mother did know and know well, since she worked in her father's store from the time she could walk. In spite of her lack of formal education, my mother was an intelligent woman and always there for her family. She would later encourage my brother Elton and me to attend, and finish, school. No greater pride had she than the graduation of both of her sons.

They had been married after having known each other since she was 13 and he, 17. He was a farm boy, and she the daughter of a merchant, and the two classes would normally not have gotten together. But Hilbert Geurin from the country charmed Willie Smith from the town of Hot Springs, and they knew from the moment they met at a barnyard dance that somewhere, somehow, they would be together.

Hot Springs, Arkansas, was a thriving place, its unique natural hot springs and cool mineral springs a pull for tourists from all over. Legend has it that warring tribes who inhabited the region long before the push westward by white settlers, and the discovery of hot springs by the Europeans, would lay down their arms to bathe in the hot or mineral springs. In the early 1800's, the United States acquired the rights to the springs. Around the springs, the spas and the mineral waters, the town of Hot Springs emerged, and much later my grandfather's general store became a part of the downtown culture.

World War I, the war to end all wars, was going strong around then, and more and more countries became involved in the dispute. Shortly after their marriage was when my father was drafted into the Army. He never served overseas because the Armistice for World War I was signed on 11 November 1918, shortly after he joined. If he had been disappointed that he never went overseas and contributed to the victory, he never talked of it. The War was over; life went on.

Life in Arkansas was a struggle. The farm didn't produce enough to support a wife. With the end of the War came an excess of men looking for work. With jobs few and far between, Dad decided to try his fortune elsewhere. Like with so many others of that time, the West called, and so he went to California.

In those days, there were no freeways, no toll-ways or expressways. There was not even a wide, four-laned highway. There was the Old Highway 66 with all of its bumps and curves and narrow ways. It followed the lay of the land, winding around mountains, and stretching across flatlands and deserts. Already, unique tourist attractions begin to spring up along this main artery from East to West. The famous highway was the brainchild of businessmen from Illinois to Oklahoma who wanted an intercontinental highway to take advantage of the new transportation rage—the automobile. The 2,400 mile long route was not completely paved when my father picked up and moved West from the Arkansas mountains to California.

The journey Westward was not easy in an old 1920 Chevrolet. He chugged along at a speedy 40 miles an hour. The windshield wipers were large and noisy and had to be hand-cranked to start. The headlights bulged like eyes of insects and had a short range. The huge high curved fenders were designed to deflect mud, but sometimes they only managed to sling the mud back on the driver. Windows were usually left down and as a result multitudes of insects that missed dying on the windshield smacked into the occupants of the car. The radiator had to be refilled constantly. Dad had two canvas water bags hung on the front bumper for just that purpose, and another one for drinking water. The drinking got hot and was unpleasant to drink, but it was all he had.

It was a week's long journey, with no shelter from intermittent storms or the heat from the desert. He didn't have the money for a hotel, and roadside motels had not yet sprung into existence. There were autocamps along the way where a traveler could pitch his tent for the night, but Dad was a little leery about staying in them because of Mom. So they slept in the car, covered by hand-woven blankets Mom had brought from Arkansas.

Taverns and inns provided food and rest for travelers, but Dad was traveling on a strict budget and most of their meals were taken at the roadside ... sandwiches made from inexpensive cuts of meat from the local butchers, and day old bread from bakers. They ate sitting on the running board of the car, or, when it was available, on the grassy banks along the road. There were no coolers, no way of keeping food cold and safe for long distances. They ate where they could and what they could as they traveled to California.

Water didn't come in plastic bottles back then, and so Dad had to fill and refill his water bags along the way from roadside springs or a friendly café, where he would get coffee for both of them and put in a large thermos bottle. They had tin cups to share. I wonder what he would have thought of his sons taking a far longer journey two decades later, not over a highway, but far away over the ocean.

An experienced truck driver, my father looked for comparable work and, after trying several smaller lines, in 1921, the same year my brother Elton Edison was born, Dad worked as a conductor for the Red Car, a commuter line which ran from Los Angeles to Long Beach, California, a distance of roughly twenty-five miles.

CHAPTER 3
California

"Incoming!" The shout wasn't panicked. It was all that more startling for that matter-of-fact way it was delivered. One of the Marines wedged in the front in the amtrack yelled mere seconds before the high hum alerted the rest of us. Instinctively, we ducked, even though it would not have mattered had the hot round found its target. The round whizzed over us to hit with a hard splash behind us. It was the first real hint of the dangers of war, that zinging-pass-us missile. We popped back up, almost as one, as the amtrack bounced its way through the flack toward the waiting black shore of Iwo Jima—Sulphur Island. It was 19 February 1945, and suddenly I was sweating in my full gear. I tightened my grip on my section of the radio I would carry to shore.

Arkansas to Iwo Jima, was a journey of more than just miles. What was an Arkansas boy transplanted to California doing in the middle of the Pacific? What crossroad decided this fate, to turn down the road that led to going toward the unknown? Choices, always choices, and an eagerness to serve, to be a part of a greater picture. It was another choice that labeled me an Arkansas native instead of a Californian. A long time ago, and far removed, from that amtrack bouncing on the same ocean that washed the shore of California.

Mom had gone to California because Dad had seen it as a golden opportunity to do better than the deadend farming or working for her father in a general store. Mom didn't like the Californian way, and felt completely out of place there. All of her friends and family were still in Arkansas, and she pined for them. Finally, when he saw how unhappy she was, they made the reverse journey and returned to Arkansas in 1923.

He was determined not to back to farming, and he didn't see himself as a store clerk. He started running his own truck, having it finance by his father-in-law. Eventually wound up owning two trucks, and paid my grandfather back for the original loan. Life seemed going his way at last.

By two years later, with my birth in 1925, Dad was struggling to support his growing family. In 1927, my father was slowly edged out of his business by the bigger trucking companies. When it became apparent that trucking would no longer be profitable, it didn't take much for his brother, Albin Geurin, to talk him into the move back to California. Mom still did not want to go, but this time he gave her an ultimatum for he did not see any way to support his family in Arkansas after the trucks were sold at a loss. They never raised their voices to each other, but Dad had a way of making it known he meant what he said. As much as he loved Mom, he was just as determined to be in a place where he could be the bread-winner. He had seen enough poverty in his life that he did not want his sons to experience the same thing. He sold everything once again, and with his wife and six- and a two-year-old sons, he began the trek back to California.

Uncle Albin lived in Wasco, California, twenty-eight miles west of Bakersfield. My Dad moved us twenty miles west of there to Lost Hills, which was forty-five miles northwest of Bakersfield. My uncle was a California State Highway foreman and was responsible for hiring and firing, and promised my father a job if he moved to California. Uncle Albin's section of responsibility was from Wasco to Paso Robles, which was rather a large area 85 miles west.

It was a long journey from California to Somewhere In The Pacific, a journey far from anyone's imagination in 1927. Who could have foreseen the chaos in the Pacific, the determination of foes so resolute that it would take the greatest fleet in the world to finally defeat them? The War to End All Wars had been over for over two decades. There wasn't supposed to be another, and certainly not one in which his two young boys from Arkansas would be so personally involved.

We packed up again and moved when I was two years old to Lost Hills, California, where my father operated a Motor Patrol Grader for repairing paved highways and shoulders. It was hot, tiring work, but he was in trucking again, and that was what mattered to him. It was work he knew and gave him a sense of independence. He would never have made it working in an office and reporting to a supervisor on a daily basis. He liked the freedom truck driving gave him.

Life in Lost Hills . . . looking back, I wonder how anyone would have deliberately chosen that remote, deserted area in which to live. Even its name was an oxymoron . . . no hills in Lost Hills. The running joke was that the hills were truly "lost." The "town," if one could generously call it that, was one general store, a gas station and a deli at a crossroads. It was the main road to nowhere and everywhere. The land was dry and arid. Dust got in every crack of the house, and in the air so it was breathed and got in the lungs. I was far away from any city. Dad also said it was far away from crime. It was also affordable, and I know that, for him, it was a combination of all three main factors that took him to that out-of-way place.

There was one bright spot in being in Lost Hills. That came in the form of a little lost, mix-breed dog. My Dad found him one day when he was on the way home from his job. He heard a mewing sound and stopped to look. He thought it had to be a stray cat, but there down in the ditch was a little wet dog, about a year old. He picked up the shivering animal and brought him home.

From the beginning, the little dog, whom we named Risty, loved to ride. Dad had a 1922 Jewett automobile. The Jewetts were a short-lived model made by Paige Motors in Detroit for only five years, between 1922 and 1927. The car had wide running boards and high fenders and iron bumpers. He wasn't trained, but whenever Dad would get in the car, Risty would jump on the running board, stretch out and ride there.

Mom and Dad never left us boys with babysitters, so we went with them everywhere. When we'd get in the old Jewett, here would come Risty and take his place on the running board. One day, Dad took a curve in old, dirt road, and Risty slipped off the fender. Risty always rode on the driver's side of the car. When he fell, Dad noticed, but it took him several seconds to pull over and stop. Risty had landed on his feet, and Elton and I poked our heads out the window and hollered, "Come on, Risty! Come on, Boy!" He came running up and hopped onto the running board as though nothing had happened.

Without Risty, I think the time at Lost Hills would have been a lot worse for both Elton and me. He was some dog! I had him all the time we were in Lost Hills, and he was the bright spot in an otherwise dreary place.

Two years later, my maternal grandparents came to California from Hot Springs, Arkansas, and stayed with us in Lost Hills for about a year. Strange, what memories stay after all of these years. I remember most about my grandmother that she dipped snuff and stirred it in her mouth along her gums with a little twig from a tree. She chewed on that twig so much that it was splintered. I swore I'd never dip snuff or smoke.

This was the time of the Great Depression, which didn't mean anything to a kid of seven. If my parents were worried, I didn't know it. I didn't know we were poor or that the whole country was in the same shape. Jobs were scarce, if at all. Businesses collapsed and without them, the economy went with them. When big companies fell, they took with them all the smaller companies that depended on them for wares and trades. Lost Hills was not the place for a man to try to do right by his family in the midst of a country-wide depression.

The economic distress led to the election of the Democrat Franklin D. Roosevelt to the presidency in late 1932. Roosevelt introduced a number of major changes in the structure of the American economy, using increased government regulation and massive public works projects to promote a recovery. But despite this active intervention, mass unemployment and economic stagnation continued, though on a somewhat reduced scale, with about 15 percent of the work force still unemployed in 1939 at the outbreak of World War II in Europe. After that, unemployment dropped rapidly as American factories were flooded with orders from overseas for armaments and munitions. The Depression ended completely soon after the United States' entry into World War II in December of 1941.

In 1935, still far from the War, we moved from Lost Hills to Visalia, California, where my father went to work for Lang Transportation Company as a truck driver, pulling gasoline trailers throughout California. The truck was an old chain-drive Mac with solid tires. Dad took Elton and me with him over the Ridge Route up the Grapevine on what is now Interstate Highway Five. That winding road got its name because its layout was as crooked as the fabled grapevines in the vineyards along the route. At that time, it was two lanes and hugged the lay of the land. It crawled and twisted and turned up the hills. We went so slow, and it was so hot in the cab of the truck, even with its snap-on curtains, Dad stood with one foot on the running board, one on the gas pedal, and steered the truck like that so he'd get some air. He drove like that all the way up the winding hill, pulling a tanker full of volatile fuel!

I was only ten years old then, but I remember so clearly how Elton got out of the truck on the other side, climbed up on the back of a watermelon truck in front of us, snatched a watermelon and brought it back to our truck. We did all of this while we were slowly moving up the Grapevine.

"Here, Knucklehead," Elton said, as he dug out a juicy part of the meat of the watermelon. I took the dripping mess and chomped into it, and the juice oozed down my chin. I wiped my face with the back of my hand.

Elton laughed and bit into his part of the melon. By the time we were up the Grapevine, that watermelon was one soggy mess of rind and red juice and black seeds. Elton held up his hands to dry in the wind. I wiped mine on my tee-shirt. Dad just shook his head at us and ducked back inside the cab as he made the crest of the hill.

Half a lifetime ago, in 1935, I was on a slow-moving tanker with my father and older brother, riding up and down the Grapevine like it was a roller coaster at a local fair. There wasn't a care in the world back then, not for the ten-year-old I was. Watching my fourteen-year-old brother crack open a fresh watermelon snatched off a moving truck, and pass it around while Dad was slowly going up the curving highway was the highpoint of my life. What ten-year-old could think past that, could ever visualize that in less time than he was alive he'd be in a boat going about the same speed as the truck climbed the Grapevine? But it would not be up California hills, but toward a faraway dark beach in the midst of a battle that would become one of the major pivotal points in the war? In the amtrack, I thought, If I lived to tell about it, who would believe me? I almost smiled, thinking of when my father told tall tales to Elton and me. One day, would it be Elton and me telling of our exploits in war? It was incomprehensible back then when snatching a fresh watermelon from a slow-moving truck was the high point of the day.

My father was fond of telling stories, sometimes very tall tales. What he wouldn't tell us was about was World War I. Even though he didn't see any action, he knew plenty of veterans who did. It wasn't a topic he thought right for telling to his boys. War wasn't something likely to happen again, and it was best put to rest. There was nothing romantic about war. It wasn't glorious or heroic. Wars were fought, and won, and lessons learned. They were not fodder for stories around the fire to inspire his boys to take up arms. It wasn't that he discouraged us from serving our country. He never would have, and when the time came he wasn't the type to talk us out of it.

Dad would talk about anything else, but not the war. Often when Dad was telling us something, the only way I could confirm it was to look at my mother. She nodded or smiled to let me know whether it was a tale or a truth. They were so in tune, I don't think I ever thought about it until they were both long gone.

My father was what was known as a *boomer* truck driver. He was hotheaded and blunt and wouldn't take any putdowns, or what he even remotely considered an insult from anyone. Consequently, he hired on to many trucking outfits, and quit them just as quickly and moved on. Because of this, we moved around a lot over Southern California. His reputation as a hothead did not move with him, because, in spite of it all, he was a good driver with a good work ethic, which he passed along to both Elton and me.

CHAPTER 4
Rumors of War

We were close enough now to see the carnage in the water. Bodies of Marines floated face down, their full packs weighing down their bodies. Here and there rifles bobbed alongside the bodies. The water was a mixture between frothy red and brown. There wasn't anything clean about the sea off Iwo Jima. It was another jolt toward the reality of what we were heading toward. How had I gotten here so fast? Had it only been ten years since I had been the youngest entrepreneur in McFarland, California?

By 1934, we lived about five blocks from a big packing shed in McFarland, California, about 30 miles north of Bakersfield. The packing shed was right on the railroad, right on Highway 99. I frequently went into the packing shed and no matter what was being packed at the time—watermelons, cantaloupes, tomatoes, potatoes—I headed straight for the culls, that part of the products not suitable for market. They were bruised, or had too many bad spots, or otherwise damaged, the throwaways, destined for the dump.

"Hey, kid, you here again?" The burly wrangler remarked as I ducked into the shed out of the blistering California sun. He may have sounded growling, but he never seemed to mind my being there. He nodded toward the large tables where the product was being sorted. "Go for it, kid!" he said.

I hurried by him, went beyond the long, flat tables straight to the large iron barrels where the bruised and otherwise spoiled fruits

or vegetables were tossed. I was careful not to bother the workers
who were busily sorting the fruits and vegetables. Their hands fascinated me; they moved so fast over the produce, tossing them on
the conveyor belts or throwing them toward the waste bins without a pause. They were migrant workers who followed the crops.
Their kids went to our school during the season then moved on.

I loaded those culled fruits or vegetables into my red wagon,
and I went up and down the street, calling, "Vegetables! Come get
your vegetables! Fruit! Cheap!" I always sold whatever I had for a
nickel a piece. Sometimes, I made as much as a dollar in one day,
which was a lot of money to a kid back then. I had to fill up my
little red wagon four or five times in order to make the dollar, but
it was worth it. That wagon had creaky wheels and faded red paint,
but it was my pride and joy.

When the shed wasn't open, I dragged my push lawn mower up
and down the street. The mower, like the wagon, had seen better
days. The handle had started out green, but was then so worn the
paint was chipped and worn to the bare wood. The blades were
good, though, because I faithfully sharpened them, using a file.

I picked the houses with tallest grass, and boldly went up to the
door. "Want your lawn mowed? Fifty cents!" There weren't many
times I was turned down. Not being a good bidder, and not liking to
barter, sometimes I had to work two days on one yard for the fifty
cents. At that time, fifty cents took me to a movie, bought me a bag
of popcorn, a soft drink, and I still had a quarter left.

Around this time, rumors of war made the news, but it wasn't
something in which I was particularly interested, and certainly
not something I could even broadly imagine I would be a part
of someday. The war was over there in Europe; it didn't concern
me. The politics of war were a foreign concept to me. I was more
concerned with the price of a movie being raised from a nickel to
a dime than the Fascists quarreling over territories whose names
were just lines in a history book.

In the evening, we gathered around the radio and listened to
the news. Elton and I lay on the floor, chins propped in our cupped
hands, while the serious voices of newscasters like Edward R. Mur-
row or Paul Harvey announced the latest news and ever growing
possibility of joining the war in Europe. Dad listened intently, but to
Elton and me it was not anything interesting or important. What we

liked were the dramas that followed, like Mercury Theatre. Unlike later television dramas, with radio the action and the visualizations were as different and as exciting as the individual listening to them. I could "see" the play come alive in my mind, and what I "saw" was, at times, vastly different from what Elton "saw."

By 1936, there just wasn't any private work to be had. It was in the middle of the Great Depression, and Dad went to work for the WPA (Works Progress Administration), working on the construction of bridges, highways, and any other project he could for about a year. The WPA was a relief measure established in 1935 by executive order as the Works Progress Administration, and was redesigned in 1939 when it was transferred to the Federal Works Agency. Headed by Harry L. Hopkins, it was supplied with an initial congressional appropriation of $4,880,000,000, an unheard amount in 1939. The program offered work to the unemployed on an unprecedented scale by spending money on a wide variety of social and infrastructure programs, including highways and building construction, slum clearance, reforestation, drainage ditches, and rural rehabilitation.

The WPA was set up by President Franklin Roosevelt to get people back to work during the Great Depression and keep them off the soup lines. It turned out to be one of the busiest times for our country. President Roosevelt was a visionary on the home front. His legacy would stretch far wider than lifting people out of a Depression. I would learn to appreciate him much later in ways that were unimaginable back then. At the time, when I thought of our President, I thought of his chomping on a cigar, not his politics.

The amtrack slowed, several yards from shore of Iwo Jima. The Marines rushed over the sidest, and we Navy radiomen scrambled with them. We waded up to our waists in the churning water. Shells rained down from Mt. Surabachi, and as bad as it was, we didn't know then it would get much, much worse. The Japanese were holding back, waiting for the Fourth Wave. For us, the muddy and bloody water, the zing of the bullets passing close to us were enough for us to know we were sloshing into hell. Yet hell on earth would not break loose just yet. The cunning of General Tadamishi Kuribayashi was unprecedented. Once upon a time, not so very long ago, all I had on my mind was becoming a baker in my uncle's bakery. Now my central focus was getting on that beach and putting my part of the three-sectioned radio together, and staying alive. It was beginning to dawn on me that going on that island might be my last act in this life. Our goal was setting up shore to ship communications, and that was the only thing on my mind. I looked around for Garner and Estrada. It had been a long journey, but there I was, on Iwo Jima in the midst of the most hellfire seen in battle until then. Life changed, almost in the blink of an eye.

By 1936 we had moved once more, from McFarland. Until 1940, we lived about seven miles north of Bakersfield in a little adobe house which my folks purchased. It was the longest we stayed in one place. It was when I became interested in the bakery and could see being a bakery chef in my own bakery as my future.

The war had started in 1939 in Europe at the time, and that was the talk at school. On 27 September 1940, the Tripartite Pact, Italy, Germany and Japan joined together in what became known as the Axis Powers. Japan agreed that Germany and Italy had the right to take over Europe and establish a new order. Germany and Italy, in turn, agreed that Japan had the right to take over the Asian countries. By joining together in the Pact, they would assist one another with all political, economical, and military means if one or the other was attacked by another power. Their combined strength, it was believed, would result in a three-way dictatorship over all of Europe and Asia.

The three countries were the most powerful in Europe and had amassed large armies under their command. With the Tripartite Pact signed, the invasion of Europe and Asia began with the determination of establishing and maintaining a new order in which those three would become the masters of the continents. To that point, the emerging war in Europe had nothing to do with the United States, except as a political and supportive friend of what would become the Allied Powers. The United States was shipping war supplies and condiments to European countries that were under threat by the Axis powers. The Axis were determined to stop those shipments before they could reach their destinations, even though they came from a country not at war with them.

The war also became the central topic at home, especially between my brother Elton, four years my senior, and myself. Elton remarked, "It won't be long before we're drawn into the war."

I agreed, even though I wasn't as sure as he why it would be so. "Why do you think so? It's so far away."

Elton had been following the news more carefully than I had. He said, "Didn't you hear that the cargo ships being attacked?"

I'd heard about ships being sunk off the coast of New York, but I didn't understand what the sinking of the ships had to do with the war in Europe.

Elton and I were sitting on our front porch. He had a baseball in one hand and tossed it up and caught it, over and over again. He was thinking hard about what he'd heard. "You know, Arvy," he said, "those ships were bound for Europe with supplies and munitions, and I heard tell it was German submarines that sunk them."

Elton tossed the ball once more and this time I caught it. "Damn," I said, "I can't see how our government will put up with that without taking action."

Elton got up, took the ball from me and walked back into the house. I sat there thinking about what he'd said. It was obvious he was worried. I knew that he had registered for the draft, as required by law. That my brother might be called up and sent overseas had never occurred to me until then. I shook off the thought. For now, the war in faraway Europe, a place we only knew from history books, was as unreal as our studying the Roman Empire in school.

My friends and I were carefree that summer. We played tackle football in the sandlot in the back of our house. We played baseball and swam in the old irrigation ditch which had a weir (small wooden dam) in it. When the farmers needed water from that canal, the weir was raised and the water flowed through at full strength.

We had a neighbor who was my age by the name of Billy Jacob Jones. We called him Billy J. He was a daredevil and always the first one into the irrigation ditch, or to go up highest in a tree, or skid the farthest on a base hit.

"You know, Bub," he said while we were peeling off our shirts to go into the ditch for a cool swim, "them Germans come to California, I'll whup their asses."

I laughed. Billy J. was a sawed off stub of a boy, which is why he acted so fearless, I always thought. "Yeah, you and what army?" I teased back, and sat on the bank to take off my shorts.

Billy J. picked up a fallen limb and made like it was a sword. "Don't need no army," he bragged.

"Sure, you're going to take them on single-handed!" I joshed and prepared to jump into the irrigation ditch.

Billy J. did a little dance with the switch. He reminded me of a swashbuckler in the movies, like Tyrone Power or Basil Rathbone.

"Beat you!" I yelled and Billy J. put down the stick and ran toward the ditch.

All the talk of war, like it was something romantic and heroic, didn't mean much at all to Billy J. and me. It wasn't anything that really struck home to me, or had anything to do with me. It wasn't like we'd actually go to war ourselves. We were just young boys dreaming of being heroes.

We never wore shoes in the summertime. My feet became so tough that I could run on the hot asphalt without feeling the heat. Bakersfield was hot, hot, hot in the summer. The irrigation ditch was a welcomed relief.

As the war wore on in Europe, there was one way my friends and I became a part of the war effort for Europe, and that made us feel pretty good and important. One day, Dad came outside where Elton and I were tossing a ball back-and-forth. Billy J. was just coming across the field toward us. Dad waited until Billy J. was there. "Boys," he said, "I have something to tell you. You too, Billy J."

Elton tossed the ball down, and he and I waited because when Dad had that serious tone, we knew he had something to say that was important. Billy J. had reached us by then, and he took one look at Elton and me and didn't even ask his usual, "Whussup, Bub?"

"There's a job I want you all to do. Listen. I'll explain."

Elton and I nodded and waited, sure it had to be chopping wood or cleaning out hen houses, or something else just as awful.

"You know there's a war going on in Europe." We nodded in unison. "The President has asked all of us to help, right down to the smallest child. There's a need for iron, copper, and glass. You kids, you take the wagon and gunny sacks, and you see how much of it you can gather, from vacant lots and the railroad tracks." He looked at each of us, one at a time. "Whoever picks up the most, gets two helpings of your Ma's banana crème pudding."

Mom's banana crème pudding was made with real bananas, crème pudding, and layered in-between with vanilla wafers. It was the best ever! For a helping of her pudding, I'd do practically anything. "Yessir!" I said quickly, and grinned inwardly because the wagon was mine. Elton would have to drag a gunny sack around.

Dad told us to bring our bounty back to the house and he would take it to the distribution stations. I felt suddenly very important as I walked up and down the streets, along the railroad tracks, and through the alleys and gathered iron, copper and bottles and brought them back to Dad so it could be used to help the fight against Germany. It was something that made me feel like I was part of world affairs, yet it didn't mean much more personally than if I had been selling magazine subscriptions door-to-door for the Boy Scouts. War was in the abstract, a game played by adults in countries I had only read about. Their casualties were just numbers reported in a paper I hardly read, not any more real than the villains shot on the street in the western movies Elton and I saw occasionally. The important thing there was the banana crème pudding.

In 1940, we vacationed in Hot Springs, Arkansas, to visit our grandmother. What did we know, Elton and I? I was fifteen then, and Elton was 19. We enjoyed being on the farm, and my mother wanted to stay and be with our maternal grandmother. My grandfather had died the previous year, and my grandmother went to live with her son, my Uncle Arvy. Uncle Arvy had nine kids, but still took his mother in to live with him. So it was that Dad and Mom went back to California, sold our little adobe home, and came back to Hot Springs. We ended up staying only a year. Visiting the farm and living on it, we soon realized, were two distinctively different things.

We didn't actually live in Hot Springs that dreadful year. Instead, we were fifteen miles outside the limits of a little community called Magnet Cove. There was just a general store and service station there. It made Lost Hills, California, seem a metropolis by comparison. We lived about one-half mile from a dirt road in a little shack up from the main house where my aunt and uncle lived. We had no running water, no electricity, no gas, no indoor plumbing. Wash water and bath water were lugged up from the creek, which was down a steep slope behind our little shack of a house. Drinking water was carried from a spring a quarter of mile away. Mother washed clothes in a great big iron tub in the backyard. She scrubbed them hard on a scrubbing board and wrung out the clothes by hand to hang them over a clothesline in the yard.

We had two mares, one wild-mannered and one tame. Dad took

the wild one to the creek one time to try to make it drink water. It bolted and stepped on Dad's leg, breaking it. He ended up on crutches. He was not fond of that mare.

We did manage to have some other farm animals. There were two pigs and two milk cows, and chickens, besides the mares and an old mule. There was no grazing land for the animals. It was not what one would call a prosperous farm, and we barely made a living. It was obvious to Dad that it had been a mistake to come back here. He begin to make plans to return to California.

One day when I was walking back to our house from feeding the chickens, I overheard Elton arguing with Dad. "I'm old enough! I am going to hitchhike back to California!"

Dad was obviously red-faced and upset, but he kept his temper. "You will stay here with the rest of the family," he gritted. "We will go back together. We pull together around here, young man."

Elton turned on his heels and brushed by me. I hurried after him. He didn't say much, but it was evident that Elton definitely didn't like it, and felt he had the right to return West. But the respect our parents taught us was stronger than his desire to be out of Magnet Cove, Arkansas.

Finding that farming the rocky ground of his in-law's acreage was not going to support his family, Dad went to work for the titanium mines. Titanium is a lightweight metal that was used as the white pigment for paint, paper and plastics. Its more important use in those years was as an alloy for parts of aircraft and submarine engines. Titanium was classified as a strategically important metal, and one of the richest deposits was in Magnet Cove. Dad drove a dump trunk in a round robin from the steam shovels scooping out the metal from an open-pit to the processing plant and back again. He worked nine hours a day, six days a week, for thirty cents per hour.

For the last six months we were there in Arkansas, Elton also drove a dump trunk for the titanium mines, doing the same work for the same pay as Dad. Every penny he earned went into a Mason jar on the dresser in our parent's bedroom for the promised journey back to California. It was hard work for both of them, but I never heard them complain. Elton was that determined to get back to California.

Talk of the war had faded into the background. News from Europe still played out on the radio and in the papers, but it seemed less important to a man determined on making, not only a living, but getting enough together to get his family back to California. Hard work, forcing a living from rocky ground, took up all our energy and thoughts. That, and plotting ways to return to California. We may have been born in Arkansas, but neither Elton nor I considered ourselves Arkansawers. I was too young to work in the titanium mines in any capacity, but there were other ways to do my share.

I hitchhiked every Saturday into Hot Springs to work in my Uncle Arvy's printing shop. My Uncle published the hotel tourist guide for Hot Springs, among other things. He paid me ten cents an hour to help him. Over a period of a year and half, I managed to save ten dollars to lend to my father in order to help get us back to California. Among us—my father, my brother and I—we saved enough for the trip. I was never so happy to get out of a place in all my life! I hated it there, and was so glad to go back in California.

In April of 1941, we finally moved back. Although Highway 66 was improved over the first journey West my father had taken, it still was only a two-laned highway, and there were parts that were poorly paved and some still unpaved. There were more tourist stops along the way, and some unique and weird buildings designed to catch the tourist trade. But we weren't tourists, and we couldn't have afforded the side-attractions anyway.

It took four days and three nights. Elton and I slept outside near the car, while Mom and Dad used the car seats for their beds. There were bench seats in both the front and back of the car, so it wasn't too bad. Dad was too tall to stretch out comfortably, but he made do with what we had. Although Elton was a capable driver, Dad didn't want either of them to drive at night. The road was difficult enough in the daylight, and at night there were places that had potholes that could not be seen and could almost swallow a car.

Very rudimentary motels were beginning to spring up along Highway 66 by 1941, some of them theme-based to attract tourists and beat their competition. There was a boon of coffee shops and cafes along the route as America was on the move by 1941 and Highway 66 was the main highway intersecting the country. The autocamps that had been prevalent when Dad first came West were giving way to more formalized and commercialized campsites, and these would eventually give way to road side rest stops with outhouses and hot coffee stands.

Our trek West didn't include stopping at any of the enticing
tourist attractions. We needed to be sure that what money we had
stretched far enough to get us to our destination.

We returned to Bakersfield, and moved in with Uncle Vernon
and Aunt Lula Smith and their sons, Jimmy and V.E. Jr. My cousin
V.E. and I were close, even though he was two years older than I.
All of us boys, except Elton, got up at 5:00 a.m. every day and made
doughnuts, fresh bread, and rolls and worked with Uncle Vernon in
his bakery shop that was in front of his house, until it was time to
go to school. This was it! This was what I was destined to do. The
aroma of doughnuts and fresh baked bread was like nothing else
on earth! People would always need to eat, and doughnuts and coffee would always be a stable. There was no way I could go wrong
by studying to be a baker. How many times had I heard that bread
is the staff of life, and here I was learning how to make it, a most
essential part of every household. Definitely being a baker was my
future!

I was now a junior in high school, and enjoyed working part-
time in Uncle Vernon's bakery. My Uncle's largest account was
making and delivering doughnuts and rolls to the schools. My main
concern back then was getting those doughnuts off to the schools,
and planning to one day open my own bakery. There wasn't a better or more studious student than I was when it came to learning
about being a bakery chef. If I had spent half as much energy on
my schoolwork, I would have graduated valedictorian!

It was there where V.E. and I planned on someday opening our
own bakery, even as we scarffed down hollowed-out loaves of
steaming hot bread, and iced donuts and sprinkled them with colorful candy sprinkles. Life couldn't be better!

CHAPTER 5
War

As I went over the side of the amtrack into the cluttered waters off Iwo Jima, doughnuts and fresh bread were nowhere in my thoughts. In the knee-high rushing waters of the Pacific as I waded toward shore, my mind was filled with thoughts of duty and survival. Even though the full force of the Japanese wrath had yet to be unleashed, the lethal rain was all around me. A Marine fell right in front of me, the splash of his body hitting the water seemingly louder than the hellfire that killed him. There was no time to think of anything but getting ashore and doing what I had prepared for all these months. My section of the radio was slung over my shoulder in its waterproof protective covering and I kept my eyes on the shoreline. There was a time, not so very long ago, that the war in Europe was still constrained to Europe and the politicians in America swore we would not go to war; yet I was in the Pacific going toward the hellfire of war.

The war in Europe continued to make daily headlines. The casualties and destruction mounted each day. Rumors of unspeakable atrocities were spoken in whispers when the adults didn't know we were listening. At night in the room we shared, Elton and I talked about the war in Europe and the unimaginable things that reportedly were going on over there. "They torture POWs," Elton said as we prepared to go to bed.

"There're laws against that," I protested, stripping down to my briefs.

Elton shrugged and climbed the railing to the top bunk. "They don't recognize that law, Arv," he said. "They're making their own law."

"I dunno," I argued as I slipped into the lower bunk, "seems to me even in war, there's got to be some rules."

"You'd think," Elton said from above me. "Go to sleep, Bud. No use talking about things that don't concern us."

He was right. So far, except for the shipping of supplies overseas, the war in Europe did not involve the United States. The American politicians were at odds, some wanting the United States to be isolationists. The Neutrality Acts were passed in 1935, which created an embargo on all war item shipments. When England and France declared war on Germany on 20 September 1939, England turned to her old friend and ally, the United States, for help with supplies. With Hitler continuing his expansion in Europe, the Neutrality Act was relaxed, and final break in isolationism began with the Lend Lease Act of 1941. America began shipping war supplies to England. The only concession America made was that such sale of arms had to be on a cash and carry basis.

The war in Europe didn't threaten our shores, said the isolationists. It had nothing to do with us, the isolationists stated, firmly and loudly. We should stay neutral, stay out of it, and let the Europeans handle their own problems. This in spite of the fact that merchant ships sent out from the United States were being sunk by the Germans.

The others, known as "hawks," pointed out that the English were our allies and London was getting pounded. The London Blitz, as it became known, was a continuous bombing of London by the German Air Force, the Luftwaffe. It was aimed at destroying the Royal Air Force (RAF), as well as other strategic targets such as aircraft factories and other industry which could be retooled for war. The Blitz did not discriminate between military and civilian targets when the bombs fell on heavily populated London, England. In one year, the German bombs killed 40,000 men, women and children in England.

The first air raids over London were aimed at the docks and industrial areas. Soon, the German bombers were dropping their bombs anywhere in the metropolitan areas of England.

America helped where we could, but felt smugly safe way across the ocean from the battles over territories, expansion and political power. It was so far away it was only fodder for movies and worry for old men, veterans of another war that was never supposed to happen again.

The war hadn't reached our shores and it wasn't likely to, so why should America become involved? What I didn't know then was that at that very moment German submarines were off the coast of New York. They were able to spot our cargo and passenger ships leaving New York Harbor because the need for a blackout hadn't yet been taken seriously. The Eastern seaboard was lit up like the Fourth of July. Ships sailed under full steam and with their running lights illuminating them. Just outside our territory in international waters, those ships became prey for a hungry predator.

In Texas, off the Gulf Coast, there were also German subs sitting out there, and the citizenry even went out there and watched them. I didn't know all of this back then, and it wasn't something that the newspapers expounded on. Reporters as a whole, and newspaper editors collectively, respected the need to keep certain information from leaking out and alerting our enemies of what we knew and how we knew it. As far as I knew, the war was far, far away.

There was growing concern that America would become more directly involved and her young men would march off to a war that, essentially—if one could believe certain politicians at the time—had to do with European borders and expansion. We didn't know then about the terrible holocaust taking place in Germany. Whispers and innuendoes of terrible things happening far beyond the battlefields were not taken seriously for it was beyond our comprehension. Such things did not happen in a civilized world, and Europe was the center of civilization.

Summer turned into Fall and headed toward Winter with the hot debates about the overseas conflict a daily occurrence. I was getting used to men stopping by the bakery, engaging my Uncle in a discussion about the War in Europe. I'd listen while pounding dough for bread, or taking the doughnuts out of the fryer.

That previous August there was a joint declaration between Great Britain and the United States called the Atlantic Charter. At the signing of the Charter, the United States built a base on Greenland, with the expressed purpose of providing protection for Britain. The Battle of the Atlantic began in earnest with German U-Boats wreaking havoc on ships at sea, and would continue throughout the length of the war.

Uncle Vernon was growing more and more concerned for his son, my cousin, V.E. who was 18 at that time. He knew V.E. would be drafted if America went to war. He was less concerned about Jimmy because of his age. I listened as I punched the dough and I shoved freshly risen bread into the oven to bake. The conversations sometimes became as heated as the bakery ovens.

There was a regular customer named Oscar Bevins whose son Henry was a year older than V.E. Mr. Bevins stopped in the bakery every morning, more to shoot the breeze with my uncle than for the doughnuts and coffee he always had. That morning, he told my uncle, "Henry up and joined the Marines." He set his hot cup of coffee down on the counter.

Uncle Vernon slung a dish towel over his shoulder, grabbed the coffee pot off the burner, and topped off Mr. Bevin's cup. "War's coming close, Oscar," my uncle said. "Your boy's liable to be in the thick of it."

Mr. Bevins nodded, wiped his mouth with the back of his hand, set his cup down and said, "He didn't want to be drafted in the Army, Vern. Saw it coming."

"Yep, Oscar," Uncle Vernon agreed as he wiped the wet spot from the cup from the counter. "Fraid my boy V.E.'s going to be drafted if this damn war gets heated up any more."

"Likely," Mr. Bevins said. "Likely indeed, Vern." He nodded at me, handed his cup to Uncle Vernon and left.

Uncle Vernon was thoughtful for a long time after Mr. Bevins left. He then looked at me as I was taking the hot loaves out of the oven, and said, "Arv, when the war comes, and it will, you take care, you hear?"

Uncle Vernon was not the demonstrative type so I knew he had to be worried. He knew far more about the goings on in Europe than I did, and he had two sons and two nephews whom he knew would be affected if worse came to worse.

It wasn't long after that those fears came roaring to life, like thunder roaring over the skies from Hawaii to California. Worse most definitely came to worse.

On Sunday morning, 7 December 1941, bombs fell on sovereign

United States territory for the first time since the War of 1812. The United States had always gone to war; war had not come to its shores since the Spanish-American War.

The Japanese settled the speculation about whether or not our country would be drawn into war by bombing Pearl Harbor. It had happened just after dawn. It was a sunny day in Hawaii and the base was, for the most part, quiet, tending to Sunday services. The ships in port, lined up side-by-side like sitting ducks in a barrel, were manned by minimal crews. There was no warning, no sirens blaring, when the Japanese planes came zooming in and all hell broke loose.

While I was just rising, Japanese planes dropped their bombs on Pearl Harbor. When I came into the front room, Mom and Dad were already listening to the static filled broadcast announcing the unprovoked bombing in Hawaii. My parents sat, close to the big old RCA Victor radio, and Elton and I listened along with them.

President Franklin Roosevelt's somber voice told of the attack, of how the United States was now at war with Japan! "My fellow Americans, December 7, 1941, a day that will live forever in infamy, the United States of America was suddenly and deliberately attacked by naval and air forces of the Empire of Japan."

It was unreal. It was then he spoke the words that would be remembered forever: "The United States is at war with Japan."

It's hard to describe the mixed feelings that enveloped me. Elton, I can see now all these years later, understood far better than I. He was already twenty, and there was no question where his immediate future lay. The worry of Uncle Vernon that his oldest son would be drafted was no longer in the abstract.

Elton and I were stunned. All the talk of war overseas in Europe, and it had come home to our doorsteps. My reaction, after the first flush of hot anger swept through me was to go get them! Wipe them off the map! Then the questions came. "Why? How? How bad?" Although directed at Mom and Dad, the questions were rhetorical.

They were grim, Mom and Dad. Dad shook his head, not so much because he did not know the answers but because he knew all too well the reality of war. "bad enough," he said, "Bad enough." He knew the war would become personal; his sons and his nephews would go. He didn't say it, but he had to know. It was in his eyes when the last echo of the president's speech faded, but his hand didn't shake as he turned the dial and the radio went silent. There followed an unnatural silence where we all stayed where we were, like statues in a tableau.

Dad got up and went into the kitchen. That broke the spell, and Elton and I looked at each other, and then went outside for a deep breath of the chilly winter air. After all of the talk about the war in Europe, we suddenly found ourselves with nothing to say to each other. We would talk, and talk a lot, later, but then it was too new, too raw, too outrageous for us to say anything. The bombs fell; and our people died, and not half a world away, but in Hawaii. Nothing bad was ever supposed to happen in Hawaii.

Four days later, Elton and I were still talking about the outrage of the bombing of Pearl Harbor, Hawaii. More details had emerged, and we knew of the Naval warships that had been destroyed. The number of casualties were just being broadcast, and the news was horrific. We both entered the front room together that Thursday, 11 December 1941, our conversation heated as usual, when Dad sharply said, "Quiet! Both of you!" The voice on the radio seemed extraordinarily loud and grim.

"Today, Germany has ceased diplomatic relations, and considers herself as being in a state of war with the United States of America."

The disembodied voice on the radio went on to talk about how Germany and Italy in unison, and as partners, declared that the United States had violated "in the most flagrant manner" the rules of neutrality. Dad leaned forward toward the radio. We were at war on two fronts. Both Adolf Hitler of Germany and Benito Musolini of Italy believed that we would be weakened by our forced fight with Japan and we would not be up to splitting our forces. They viewed America as reluctant to fight. It was the perfect time for them to "take down the imperialists." The call for isolationism had not been lost on those dictators. They saw an easy victory. What neither Hitler nor Mussolini understood was that when Americans were threatened, they stood together, not apart. Mussolini in Italy believed, like Hitler, that the United States was too weak to be spread over three fronts.

From the balcony over the Piazza Venezia in Rome, Italy, Benito Mussolini, dictator of Italy, pledged that the "powers of the pact of steel" were determined to win. The Axis Powers were born.

The United States was not considered a major power back then, and her armed services were ranked seventeenth in the world! It was not seen as a mighty force by her enemies. What Japan, Germany and Italy did not see was the might of the American people. Even the sworn isolationists were shaken to the core by the bombing of Pearl Harbor and the reports of casualties and ships destroyed by what was becoming known as the sneak attack.

President Roosevelt, who, just four days earlier had announced the unprecedented attack on American soil, was back on the radio grimly, but firmly, telling America of the pact among Germany and Italy against the United States. "The forces endeavoring to enslave the entire world now are moving towards this hemisphere. Delay invites danger. Rapid and united efforts by all peoples of the world who are determined to remain free will ensure world victory for the forces of justice and righteousness over the forces of slavery and barbarism"

The United States was now part of World War II. It was *TIME Magazine* who had dubbed the war in Europe as "World War II" in an article two years before in September of 1939. The war wasn't political; it crossed both party lines. All American lives were threatened. It was a matter of survival for the United States. Although there were still fervent isolationalists, members of Congress did very little quibbling, and a new law was swiftly passed in Congress that allowed United States servicemen to fight "anywhere in the world." That edict all but ensured that Elton and I, along with our cousins V.E. and Jimmy, would be in the middle of the conflict, and soon.

The double whammy of the shock of the attack on Pearl Harbor coupled with the declaration of war against the United States by the Axis Powers had the opposite effect on American citizens then expected by our enemies. If they thought that Americans would withdraw in fear, they were deadly wrong. Instead of being cowed, there was a sudden and overwhelming patriotism and a surge to volunteers for the US Navy, Merchant Marines and Marine Corps, which were not part of the draft. The Army grew to capacity as a result of both the draft and the eager young men who were anxious to defend our country.

There was no doubt that Elton and I would join that surge. The only questions were when and what branch. Dad was resigned; his sons were destined to be a part of it.

At night, in the room we shared, Elton climbed onto the top bunk and said, "Bub, I don't want to be a foot soldier."

I threw back the old quilt and got into bed. "I sure don't blame you there," I said, and couldn't imagine my brother sloshing through mud in some faraway land. "Think you'll go?" I asked as I snuggled under the covers.

The upper mattress bulged with Elton's weight as he turned over to get ready for sleep. "Don't think I'll have a choice," he said.

"What'll you do?" I asked.

"I'm going to join, Bub, soon as I can. The Air Corps." His voice drifted off.

"The Air Corps?" I asked, astonished. I thought he meant he was going to fly, and I knew that only officers were qualified to be pilots. "What'll you do there?"

"I don't know, Bub," he said, and I heard him punch his feather pillow to get comfortable. "I'm good at my hands. Bound to be something I can do."

"I'm going to join the Marines," I said, confidently.

"Good luck, Bub," Elton said. "Now go to sleep."

I lay there for quite a while thinking over what Elton said before I drifted off to sleep. My last conscious thought was my own determination to join the Marines.

It wasn't long before it was evident that the war had not only gone to Pearl Harbor, but was incredibly right off our own shores.

That grim winter of 1941 rolled into January 1942, but there were no celebrations, no parties proclaiming this to be the best year ever. There were, instead, young men lined up at recruiting stations, waiting outside on sidewalks, eager to join. The War had begun!

Hitler became serious about the submarines off our coasts. Back then they were called either "underwater boats" or "U-boats." He now ordered them to sink any ship going out of the harbors, especially the oil tankers from Texas. Three days after Pearl Harbor was bombed the merchant ship *Emido*, a tanker, was torpedoed and its crew of five were lost just off the west coast.

No ship was safe, especially the cargo vessels out of the New York Harbor. Blackouts were now mandatory along the coast, but for some reason, New York did not take the matter as seriously as it should. Ships leaving New York Harbor were silhouetted against the lights of New York and were easy targets for the German U-boats. It took a while, but finally New York was blacked out, at least along the harbor.

Almost one month to the day after Pearl Harbor was bombed, on 8 January 1942, the United States Army Transport Ship *General Richard Arnold,* which was a mine planter headed overseas, was sunk off the east coast. A crew of ten was lost.

It wasn't only military ships which fell victim to the German undersea boats. In the course of two weeks, thirteen merchant ships were either sunk or heavily damaged off the east coast of the United States. The loss of life was staggering: over 225 men went down with the merchant ships during that period. These were freighters and tankers, which were torpedoed and shelled by an enemy sitting just off our coastal waters, lying in wait for them. Neither did other merchant ships going across the Atlantic or going out through the Gulf of Mexico escape the prowling U-boats. Even ships leaving the ports in Alaska were attacked. The U-boats had widely deployed, and their captains were well skilled in strike-and-run technique. This carnage would continue for another six months.

At this time, because the harbors were so busy and the makeup of the crews crossed all ethnic groups, especially German and Italian, loose talk about ships and their destinations flowed like water over a dam. This was a concern for the safety of the ships and their crews, and soon the saying "**Loose Lips Sink Ships**" became more than just a slogan on a war poster. It was literally a matter of life or death, and taken very seriously.

Now that we were at war at both fronts, there was no longer any question that both V.E. and Elton would be called to duty. Neither one wanted to be drafted into the Army and be foot soldiers, but both wanted to serve and knew if they did not volunteer, they would be drafted.

V.E. and I talked about enlisting. I was still too young, but V.E. knew it was just a matter of time before the choice would be taken out of his hands.

That morning in mid-January 1942, V.E. came into the bakery and did not don his usual white apron. He walked up and down the already-hot ovens, not saying anything at first. Then he came up to me and said, "Arv, I just enlisted."

I wiped my hands on my apron and left the wooden spoon in the batter of doughnut mixture. "Does Uncle Vernon know?"

V.E. nodded. "We talked about it last night."

I swallowed hard. I knew he would go; it was just when it was finally here, it was hard to digest. "When? When do you go?"

"Today," he said. "Have to report today."

There was no waiting between time of sign-up and time of going off to basic. Men were needed too badly. I nodded and stuck out my hand. V.E. took it and shook it hard, then pulled me in for a bear hug. That was unusual, because although we were close, he, like his father, was not demonstrative. He thumped me on the back then stepped back. "Well," he said, "see you around, cuz."

"Yeah," I answered. "See you around, V.E."

When he left, it was unusually quiet in the bakery. The fire in the ovens still roared; the chatter from up front where customers dropped in for their daily coffee or box of doughnuts still floated back to the kitchen. Yet, without V.E. there joshing with me, snatching hot loaves from the oven, it seemed way too quiet.

With V.E. gone, I knew that Elton wouldn't be far behind. That day came all too soon in February 1942. Elton, at twenty, not wanting to be drafted and conscripted into the ground forces of the Army joined the Army Air Corps, as he had told me he wanted to do. There was no United States Air Force back then. The Air Corps was an elite part of the Army. Elton wasn't a pilot, but he knew he could do just as important work, maybe even moreso. He passed the rigorous tests to become a maintenance man on the planes.

With Elton's joining, the war became even more real and personal. As proud as I was of my older brother, I knew it wouldn't be long before I, too, followed in his footsteps. The war invoked patriotism that had lain dormant since World War I, which I only knew from dusty history books. This was real; this concerned my brother, my cousin, my friends, and, would ultimately involve myself and my younger cousin Jimmy, V.E.'s brother. War was no longer in the abstract. It was up-front and personal.

MovieTone Newsreels at the local theaters were full of images from Pearl Harbor, the Pacific, and from the European Theater. The voices of the reporters mirrored the horror and, yes, the excitement, of war. There were no images of wounded or killed soldiers; those types of pictures weren't news fodder back then. But the burned out and bombed out buildings of Europe, the hulls of sunken ships at Pearl Harbor, brought to life the reality of war. In each of the images I imagined I could see Elton, my brother, and I wondered every day what it was really like over there.

Ships continued to go down off the coasts, but these didn't make the news as often as statistics of casualties overseas, or the destruction of entire cities from incessant bombings by the Luftwaffe.

Letters from the front were few, and those we did get were read over and over again. Blacked out or cut out passages fascinated me, and I didn't truly understand censorship or the reason for it. I only knew that somewhere, "out there" was my brother and whatever he was writing was important.

Life on the home-front went on as normal as could be with both a brother and a cousin in the war now. There were drastic changes in the work force. Jobs typically held by men were now being filled by women who had never worked away from home before. Women were becoming more than secretaries and nurses.

The face of America had changed, with so many of her young men going off to war. No longer did the politicians talk about isolationism. Now even those who had sworn the only road to peace was to keep to ourselves were supporting the war effort. Men too old to go to war, or rated "4F" by the Draft Board, joined ranks of others in industries that retooled and were aimed toward gearing up the armed forces, in building tanks, planes and ships. Young children still gathered copper, and brass, and bottles, but now for America instead of Europe. Women gave up wearing silk stockings; the silk was needed for parachutes.

Rationing became the norm. Ration books with their stamps were needed to buy anything, especially beef, milk and gas. Luxury items weren't even a consideration. Every household had to rethink how they spent, how they fed their families, and how they worked.

American industry retooled its mighty machines for the war effort. Tanks rolled off assembly lines that had previously produced luxury automobiles. Tire plants were now cranking out tires for jeeps and transports instead of for the domestic car. Silk was reserved for parachutes instead of women's stockings and blouses. Steel, rubber, silk were all concentrated for making the tools of war.

The make up of the workers in industry changed with the war. Women, who until then mostly stayed at home, or were nurses, teachers or office workers, now filled in the gaps left by the departure of so many able-bodied men to boot camp and eventually overseas. The term "Rosie the Riveter" soon became symbolic of the changing roles of women. Everywhere, women were an important part of the support of our troops over there. Jobs that were traditionally "men only" became woman-held. They were in defense plants, making everything from bullets to bombs; they were in ship yards welding ship hulls; they helped put planes together; were central welders on bridges and other infrastructure; they dug ditches and repaired water mains. There was no talk about "women's liberation movement;" they just did what was needed without anyone's much questioning it. Women now went off to work in coveralls, and carried lunch buckets instead of purses.

Women even flew the newly built bombers from the testing fields in the United States to Hawaii and Europe. They did this in planes that were not armed, and flew through enemy fire. Jacqueline "Jackie" Cochrane, one of the few recognized women aviators of the era, was asked by General Hap Arnold to organize the Women's Flying Training Detachment to train women pilots to handle basic military flight support. His idea was to free men for the more serious duty of flying planes in battle. Cochrane led the way, and the result was a group of women who proved invaluable to the war effort. They transported planes overseas, tested various military aircraft, taught aerial navigation, and provided target towing. During their service, Cochrane and her crew of women delivered 12,650 planes and flew more than 60 million miles.

Every day, America was on fire with the ever increasing need to produce bigger, better, more accurate weapons of war. The Depression went out with a BANG! There were still some hardcore isolationists who tried to convince the American public that the war was staged, that the president knew about the targeting of Hawaii before the Japanese bombs rained down, and did nothing, just to draw America into the war and end the Depression. That rumor, although disproved and squashed, would linger in the back of some minds for years after.

Every day, more and more of my peers, my friends, my neighbors, went off to war. There was no longer talk of soldiering as romantic and somewhat cavalier. The real-life meaning of war came home every day, posted on bulletin boards outside of churches, town halls, and town centers. Casualty lists grew daily, as did the descriptions of the destruction of towns, cities, and civilians. There was not instantaneous news; no on-the-sight, or embedded, from the field reporting, and so the figures and the names and the horrors of war on those bulletin boards were sometimes days and weeks old. It was enough for anyone scouring the list to wonder if a familiar name would be added . . . tomorrow, always tomorrow.

Elton ended up in the Burma-China-Indian Theater of War. He became a crew chief on a C46. As a crew chief, Elton was responsible for the airplane maintenance, and keeping the plane in the air without mechanical problems. He flew with it, and took care of it on the ground. He even got to name it and called it "The Donna" after his wife, a Marine he had met while in training stateside. He said in one of his letters home that, "All the pilots do is fly this thing. I keep it in flight."

He was shot down over the Burma-China Hump. That was a route that was so dangerous, and there were so many crashes, it is said the sun's reflection on broken aircraft gave the valley its morbid name of "Aluminum Valley." In order to get over the 17,000-feet mountain peaks, the pilots had to fly at 25,000 feet in very adverse weather conditions. Thankfully, he and the crew bailed out and walked out of the jungle into safety. We didn't learn of it until long after he was safe.

That June 1942, we moved back to Visalia, California, where I went to work as a baker's helper during the week, and washed bakery trucks on weekends. The war was never far off in my mind, though. As eager as I was to join, at sixteen, I wasn't able to enlist in the service. I kept asking Dad to sign for me so I could join. He kept telling me to wait. Waiting was not easy. All around me, my friends were eagerly marching off to war—high school graduates one day, inducted into the service the next. They were young boys, hardly men at 17 or 18, who eagerly and gladly faced the duty that awaited them. The rigors of basic training were not even considered or discussed. It didn't matter how we got there, only that we did. We were not old enough to vote; not old enough to drink, not even old enough to get married in most states, but we were old enough to go off to war. We never questioned that quirk of fate.

I knew I wanted to join the United States Marine Corps. It seemed the most glamorous of the services. War, the reality of war, was something I heard about and prayed about, because my brother and cousins were still there. In spite of all the images on the MovieTone news, the daily casualty lists printed in the newspapers, the war was still, if not glamorous, at the very least the ultimate goal back then to a seventeen-year-old. After all, at that age, I was invincible, and so it was the Marines that I was determined to join.

For a year, I tried to talk my father into letting me join the Marines. He steadfastly refused to sign for me. It wasn't that he didn't believe in the war or the cause for which we fought, but that I was too young to make such a decision. He already had one boy off to war. He wanted me to be absolutely sure that this was what I wanted before I did something foolish. One did not join the service and then say, "Oops, made a mistake; think I'll go home now." He also knew that when I joined there would be a very real possibility I would be in the middle of the war.

My cousin, Jimmy Smith, V.E.'s younger brother who was my age, joined the Navy when he wasn't yet 17. He lied about his age, and his mother, my Aunt Lulu, signed for him. I wanted desperately to follow his example. Back then, the military wasn't too particular in checking ages and even identities. Warm, willing bodies was all it took. They looked the other way if a recruit seemed to be younger than the required 18.

I figured the Marines were macho, and they were the ones who would ultimately win the war. I was up in Tule Lake, California, with my uncle and aunt when I decided it was time for me to join. I wasn't yet eighteen, and so had to get my parents' permission. In September of 1943, three months shy of my eighteenth birthday when I would have to register for the draft, I finally talked Dad into signing for me.

I caught him just as he had gotten home from a long drive. He was tired, and he had other things on his mind besides my continued nagging at him. He had just put down his thermos on the kitchen counter when I pointed out to him, "Dad, if I wait until December, you know what will happen. I'll get drafted. I don't want to go in the Army. How about it?"

Dad sat down on a kitchen chair and waited while Mom poured him a fresh cup of coffee. Mom didn't say anything; she always deferred to Dad, although I'm sure they had long discussions out of the hearing of their sons. "Go on," he said.

"If I join now, I can have my choice. How about it?"

Dad looked at Mom and, while I didn't see her nod, he must have seen the answer in her eyes. Resigned or agreeable, I didn't know. All I do know is that Dad sighed and his lips were tight together as he nodded. "All right," he said, "if that's what you want."

I let out a breath I hadn't realized I was holding. "All right," I answered. "Thanks, Dad."

He lifted his cup to his lips and took a long drink.

I hugged Mom. "Don't worry," I told her. "Everything will be okay. You'll see."

She didn't look too convinced, but she didn't protest either. I knew that both of them were worried about Elton, and to have another son in the midst of battle couldn't have been too comforting.

CHAPTER 6
Enlistment

Determined to beat the draft's inevitable decision to put me in the Army, the next day, Dad's written permission in hand, I enlisted in the Marines. The first item on the agenda was the physical. There was nothing that prepared me for flunking the physical because of my eye—a lazy left eye that wouldn't track right. Years earlier, that eye had been full of mud and dirt due to a childhood accident, and while I still had sight in it, it didn't function in unison with my other eye. I was frustrated. After all, I knew I could still sight down a rifle barrel, and what else did they need? I was able-bodied, strong, and athletic. There was no reason, to my way of thinking, why I wouldn't be as good a Marine as any other.

V.E., my cousin, and Elton, my brother, were both in the Army Air Corps, but that wasn't for me! I knew if I didn't do something soon, the Army would take me, lazy eye or not! If the Army snagged me, I knew that I wouldn't be as lucky as Elton and V.E., who were both in the Army Air Corps.

The Marines turned me down, and I had to swallow my disappointment like a man. I couldn't understand their reasoning, but it wasn't something I could fight. They could afford to be choosy; they weren't lacking for applicants!

Still determined to find some way to join, and not having been classified "F" in spite of my lazy eye, I went from the Marines recruiting depot straight to the Navy's. That was one of those pivotal moments, a step in time that changed everything for me, and set me on my journey to that small black dot of an island in the Pacific.

The United States Navy decided a lazy eye didn't make a difference in my ability to serve. So it was into the Navy, not the Marines, that I eagerly went. I figured I was a shoo-in to go to Cooks and Bakers School. I was promised, upon enlisting, I would. After all, that was my skill, and the Navy had to eat. The prospect of becoming a Navy cook was an exciting opportunity for a baker's helper. It may not have been as glamorous as being a grunt in the Marines, but it was what I did best, and I actually looked forward to it. I envisioned myself being a cook in a ship's galley in the middle of the ocean. It wasn't a bad image; I'd still be in the war and contributing in a very meaningful way. Besides, the way I figured it, cooking for a ship's company would go far toward my goal of becoming a commercial baker and owning my own bakery. It wasn't that I was trying to get out of being in the middle of battles; it's just that was where my experience lay.

All of us new recruits were assembled in a staging station in Bakersfield. It was a large warehouse in which was nothing but a long table behind which were the Navy officers. We all were assembled in front of them. One of the men in his dress whites, a commissioned officer, stood and told us, "Raise your right hands." All of us did, and the warehouse was suddenly alive with the shuffling of feet as we stood at awkward attention. In the quiet that followed, the officer said, "Repeat after me." The officer began the oath, stopping after every sentence for us to repeat it, word for word.

I began reciting the oath: "I, Arvy Geurin, do solemnly swear that I will support and defend the Constitution of the United States against all enemies, foreign and domestic; that I will bear true faith and allegiance to the same; and that I will obey the orders of the President of the United States, and the orders of the officers appointed over me, according to regulations and the Uniform Code of Military Justice. So help me God. I swear that I am fully aware and fully understand the conditions under which I am enlisting."

After being sworn in, we stayed standing in our not-too-straight line, still in our civilian clothes, our bags at our feet, and the petty officers had us count off: **one-two, one-two**. Then the *ones* were ordered to step forward. I was a *one*!

An officer in whites was holding a metal clipboard. He stood facing us, right in the middle in front of our ranks. He said, very loudly, "You number ones will be going to San Diego Naval Training Station." Then he waved toward awaiting bus. "Get aboard now." I hurried toward the bus. The cargo bays were opened, and the bus was idling. The bus driver was waiting on the pavement, and counted us as we climbed aboard. We had been instructed to bring a single bag, and that our last name and social security number were to be written in bold, black letters on it. The cargo bays were opened, and those bags were tossed in the bowels of the bus.

While we were walking toward the bus, the same officer announced, "The rest of you will be going to Farragant, Idaho." Then he waved toward the second bus, sitting alongside the first one. Farragant, Idaho translated to me as snow and ice. I was thankful I was a *one*! Now I was a transplanted-to-California-boy, who knew what it was like in the snow and ice and slush of Arkansas and that Idaho was much worse. I was smugly thankful for what I saw as a fortuitous selection.

Aboard the large white bus with UNITED STATES NAVY emblazoned on its side, the atmosphere was light and anticipatory. I managed to climb over one cocky recruit to get to a window seat. We were all volunteers, and were going toward an exciting, unknown adventure, to do our part in fighting a war that had come to our shores. It was a good feeling. As the time passed, the talk quieted. Some slept; others read paperback books; while still others, including myself, spent the time looking out the window at the passing and changing landscape.

It was almost 250 miles from Bakersfield to the Naval Station in San Diego in the south end of our very long state. At that time, there were no swift and easy interstates. The bus crawled over the Grapevine, weaved through mountainous terrain, paced with traffic into and out of the small and large towns, passed over the straight seemingly-endless highway across the desert, and dipped down south into the cities and finally to San Diego. There were stops for fueling, and we were allowed off for a quick restroom break, in pairs, but always under the watchful eye of the Navy officers. We were theirs now, and it was their job to make sure we arrived intact in San Diego.

On one of those fuel stops, a fellow named Nathan Olson and I stepped off the bus together, to use the restroom, mainly in order to stretch our legs. Olson, like me, was tall and the lack of adequate legroom between the seats of the bus cramped him. We chatted about what was to come.

"Been a sharpshooter all my life," Olson said as we made our brief break off the bus last as long as possible by walking slowly, "hope to get into Special Forces."

I had no idea if the Navy had Special Forces, but I was impressed. "I'm going to be a cook on board a ship," I said.

Olson said, "Hey, armies move on their stomachs." Then he laughed at mangling the old saying. "Navies, too," he hastened to say.

"My cousin V.E. and I are going to open a bakery after the war."

Olson and I started back to the bus. "I'm going to Wyoming with my brother. We're going to do cattle ranching. Heard tell it's wide open out there."

We got back aboard the bus, just seconds before the Ensign in charge reached us. We saw him coming with that "Hurry up!" look in his eye.

Olson took his seat further back and I squeezed over the legs of the fellow sitting in the aisle seat. After that bus ride, I never saw Olson again, but I often thought of him and wondered if he made it through the war and got his ranch out in Wyoming.

By the time the large bus pulled into San Diego Naval Training Station, there wasn't much talking aboard. It was a very hot day, and the windows were down on the bus. Those sitting in the window seats leaned out to take their first look at the Navy base. Those who were in the aisle seats stood up and looked out over the heads of those still sitting.

The Navy base was busy and buzzed with both foot traffic and covered trucks moving slowly away from the entrance. The guard station was a small white, wooden building with a pointed roof. A single guard in a white uniform, holding a rifle manned the station. The bus slowed on the approach. We stopped at the guard station at the main entrance, and were waved through immediately.

As the bus went slowly along familiar-to-it streets, we passed a line of uniformed recruits. They waved at the passing bus and shouted in unison, "You'll be sorry!"

I was thinking, "No, not me!" I looked around, and the same sentiment was reflected on most of the smiling faces of the recruits.

We had all worked hard to get here; looked forward to the time we would be one of those marching figures in white uniforms. There was nothing but adventure ahead; how could I be sorry for joining?

The bus weaved around the streets, leaving the recruits behind, and stopped in front of a large white building with a rusty-brown roof. The wheezing of the brakes hadn't yet died when we were ordered off the bus. We scrambled down the steps, looked around, and were told, "Line up! Eyes forward!"

It felt good to stretch my long legs. At six-one, I was one of the taller of the recruits. We stood in a somewhat ragged line. When everyone was off the bus, the driver opened the cargo doors. The bags were tossed out of the cavity of the bus onto the pavement. We were told to, "Retrieve your bags. Back in line!" The orders were barked and clipped. We hurried to. I spotted my black gym bag under a large black suitcase and snatched it, slung it over my shoulder, and got back in line.

When all of the bags were retrieved, we were ordered to pick them up and follow the petty officer into a large auditorium. We filed in, and our footsteps echoed in the tall building. There were rows of student-desk chairs, similar to what we had all left behind in our various schools. The petty officer went to the front of the room and with a downward motion of his hands indicated we were to sit. The noise level was high until we were all settled in our chairs. I dropped my bag at my feet and there were **THUMPS** as other bags fell to the floor. Then ... silence.

The petty officer looked us all over with a practiced eye, then nodded to another uniformed man, whose entrance we had not noticed, intent as we were on our surroundings.

"Do not pick up the papers that are being distributed now until you're told to." A thick stack of papers was piled on each of our desks. When they were all distributed, the uniformed aide went back to the front and stood at position rest (which we were yet to learn that was what it was called).

"In front of you," the petty officer said, "are your aptitude tests. Turn them over and put your name, last name first, on the first page."

I did as we were told and waited.

"These tests will give the Navy your strengths and will determine in what specialty you will be placed. Gentlemen, begin."

It hardly seemed to me that the tests were necessary; after all, I had already been promised Cooks and Bakers School. I was naive enough to believe that, no matter what my scoring, I would be a cook in the Navy. I bent my head to the task at hand and carefully filled in all the squares and blanks. The auditorium was quiet, with the only sound pencils being scratched across paper. The young sailor who had passed out the papers stood stock still, eyes front. I looked up now and then at him and marveled at his statute-like pose.

When I finished, I lay down my pencil and sat up straight and waited. The petty officer walked up and down between the desks, his hands clasped behind his back. He reminded me much of my high school teachers during finals. I almost grinned at the comparison ... almost. I knew better. This was far more serious than passing a grade in high school. I was smugly confident that the aptitude test was an exercise in futility for me, as there was no doubt I would be heading to Cooks and Bakers School after basic training. After all, I was a trained baker. What would they make me? A tank mechanic?

I spent the next couple minutes thinking about the difference between cooking for a crew of a Navy ship and baking doughnuts and bread in my uncle's bakery.

When everyone was finished, the petty officer nodded at the same sailor who had passed out the aptitude tests, and he walked up and down the rows, gathering them up. When he was finished and took them up front, the petty officer spoke again. "Get up. Pick up your bags. Follow me."

I pushed back out of the student desk, grabbed my athletic bag, and assembled in one long line. I was to learn that standing in line would be the order of the day. "Leave your bags here." In unison, we dropped our bags to the pavement. I did not see that bag again until I had finished basic training. The writing of my name and social security number on it was what ensured that it would be eventually returned to me.

We marched right around the corner to what was laughingly called a barber shop. "A little off the top!" was a fantasy. Here, Navy barbers stood behind a long row of identical barber chairs and, using electric shears, gave the hapless recruits a buzz cut. Now my hair was thick and wavy. I ran my fingers through it and thought,

"Oh NO!" but then mentally shrugged. Hair grows, was the only comfort. Soon it was my turn, and I sat in the chair and all I heard was the buzz-buzz of the razor zipping over my hair. When I was shorn almost bald, I joined the rest of my group outside.

When we were all assembled again, we were off to the company stores. Here it was. At last! This was the true beginning for we were being issued our first set of Navy uniforms! I was beginning to feel like this was real; I was in the Navy now!

I stood in line again. It seemed everywhere I went, there were lines, lines and more lines. Inside the company store were large bins, one stacked on top of the other, each one having a different item or size. It was much like there are at department stores, only no smiling and helpful store clerks here. Next to each bin was an unsmiling sailor who tossed or handed the items to us recruits.

The first item issued was a white sea bag. This was folded when it was tossed to me. I immediately unfolded it and hauled it along. As I went out the end of that building to another building directly across from it, and very similar to it, the petty officer in charge of us barked, "You will yell out your size, loudly! You will put your uniform in your sea bag, and move along quickly!"

Behind the long counters, sailors barked, "SIZE?" I yelled back my size, and as the line moved along I was handed a stack of uniforms, a pile of blankets, military issue boxer shorts, socks, shoes, boots. All of this, I stuffed in the sea bag. When I exited the stores, I joined the others outside, where we were grouped for marching. I slung my sea bag over my shoulders, and fell to not too gracefully, and waited for the orders to march back.

From the stores, we marched to the barracks. Outside the barracks, we were ordered to call out our name, and were lined up alphabetically, according to our last name. From that point on, I would no longer be called "Arvy," but "Geurin."

We then entered the barracks. The barracks were long, wooden buildings painted white. Inside were twin rows of two-tiered iron bunks with nothing but a thin mattress on them. At the foot of the beds were metal footlockers, the lids wide open to show they were empty. We were ordered to dump the sea bags on our beds. There were two long rows of double bunks facing each other, separated by a mere three feet of space, with small footlockers at the end. I chose the top bunk; below me was Emilo Guiterrez from San Diego;

across from me was William Garner.

As soon as the sea bags were on the beds, we were ordered back outside; where we stood in line once more and were marched off to the infirmary. "You will roll up your sleeves to bare your shoulder. You will move along the line with your elbow bent and arm across your chest. You will not talk, and you will follow all instructions given you."

I was wearing a short-sleeved tee-shirt, so rolling it up to bare my shoulder was no problem. Some of the recruits had made the mistake of wearing long-sleeved shirts, and struggled to comply with the instructions. There were a few who chose the more expedient method of taking off their shirts and tying the sleeves around their waists.

We formed an orderly line, one behind the other. This line moved slower than the previous lines I had been in that day. Smartly uniformed sailors stood behind metal tables with their hypodermic needles in hand, and boxes of fresh ones on the tables in front of them. As I moved slowly by them they jabbed the needle in my arm. Shots. Shots and more shots! Sleeve rolled up, shoulder bared, march along and get stuck and prodded. I never liked needles, but neither was I afraid of them. I was determined to get through this part without flinching . . . at least outwardly.

At the end of the inoculation line, I was told, "Sit and hold up your arm." I sat in a metal chair and held up my arm. A thick rubber band was tied securely around my upper arm. The smiling sailor smartly tapped my arm, making sure a vein was prominent. Jab! Then the blood taking. I resisted the urge to watch and stared straight ahead. After the blood draw, he unsnapped the rubber band and indicated I was to be on my way.

It seemed, before the end of the day, my arm was nothing more than a sore pincushion!

Somewhere during that orchestrated march through the first steps of basic training, a strange thing happened. It was rough, it was fast, it was slightly painful; it was just a hint of things to come, but I realized I was in the Navy. I had wanted to be in the Marines, but here I was, in the Navy, and it was a good feeling.

It was all worth it, though, to know that, at last, I was really in the service, and soon would be joining the others, the real soldiers and sailors, as soon as the training was over. The coming six weeks couldn't pass fast enough for me. I was already imagining myself in a ship's galley, making pancakes for the early morning watch.

That day, I returned with the others to the barracks. The Ensign said, "Listen up! You will make your beds, and you will put away the items in your sea bag. The petty officer will demonstrate once, and then you will do it exactly as he has shown you."

The petty officer deftly demonstrated how to make a tight bunk. Then he opened a footlocker and indicated what went where, and did the same with the metal locker. His demonstration was swift and practiced, and he gave no chance to ask questions.

The next few minutes, I made up the bunk, and put away the items that I had stuffed in the sea bag. Although the petty officer watched us, he did not demand silence. I didn't know it then, but that was to be our last truly free time until the training was over.

When we were all finished, the Ensign said, "You will stand in front of your bunks." The petty officer walked up and down the aisle between the bunks, inspecting our work. Although I was absolutely certain that I had done my best, I knew that the bunk was not nearly as tight as the one he had done. When he was finished, he returned to the front of the barracks.

"You will now shower and leave your civilian clothes. You will wear your uniform from this moment on, you belong to the Navy."

The showers were not private. They were located at the end of the barracks. There were several shower heads in the tiled shower room. I stripped and stepped in, and the lack of privacy didn't bother me because, as an athlete in high school on the swim team, the showers were much the same. When finished, I put on my first Navy uniform, which type I would wear throughout basic training. I felt, at last, like a real sailor.

Crammed in narrow barracks with forty-nine other guys, friends were made fast, friends whom I didn't know if I would see after boot camp. We were all in this together, and the commonality was a good foundation for friendships to form. The war was real; we were here for the expressed purpose of joining that conflict. It wasn't something spoken among us, but we all knew that of the fifty of us there, some of us would not be coming back. We may have felt invincible as young, eager men, but down deep, we were well aware of the growing casualty lists. Those lists were not posted at the training part of the Naval station, and the news we got was from home, not from anyone on the base.

Early that first morning, before dawn, I rose to a blast of reverie on the loud speaker in the barracks. I tumbled out of bed, yanked on the uniform of the day, and hurried with the rest of the group outside. There, in the early morning before dawn, the petty officer shouted orders. This was my first calisthenics with the Navy. I had been good at athletics in high school, and a swimming champion, so exercises were not new to me. However, I had never done anything so intense as that first day of morning exercises as a Navy recruit. Jumping jacks! Stretches! Sit-ups! Quick! Quick! The petty officer marched up and down between the rows of huffing and puffing new recruits. This was just a small taste of what was to come. Even the most athletic of us had to push ourselves to keep up the pace set by the petty officer.

Finally, we did our cool-down stretches and snapped to attention in line. Ordered to march, we did a lot better than our first march to the barracks. I knew we would get even better. We went to the mess hall. We stood outside at parade rest, while the petty officer snapped, "Inside, you will take your trays. You will fill those trays. You will eat what you put on those trays. You will not talk. When you are finished, you will exit the building and muster here."

Inside the mess hall, the noise came only from the serving lines. Sailors behind the buffet line scooped up oatmeal, scrambled eggs and hash brown potatoes and dumped them on a metal, partitioned tray as the I moved down the line. Toast was slapped on the tray. At the end of the line, I took a metal fork, knife and spoon from round metal containers. After moving through the line, I carried my tray to the first empty seat. Before I could sit, I was ordered to move on. The rule I was soon to learn was that tables were filled from the back to the front. One did not simply go to a vacant table and plop down to enjoy breakfast alone.

When I finished my breakfast ... and I ate every last bite, as I was famished ... I took the tray over to the end of the mess hall where the galley help took it from me. I hurried outside and joined the rest of my group who had finished ahead of me. When our entire company was assembled, we waited at parade rest for our instructor.

The petty officer came out and shouted, "Attention!" We all stood as straight as possible, arms down at our sides, eyes straight ahead.

"For the six weeks you are here, you will be under my command. If you survive this training, you will go to your technical school." I was grinning inwardly, thinking that six weeks would be a breeze and then off to Cooks and Bakers School. "Listen up! Step forward when your name is called." The petty officer was handed a clipboard. He glanced at it and called out names. I heard "Garner" and "Geurin" and we stepped forward. When he had finished the first group of names, he announced, "Upon your successful completion of basic training, you will be attending Radio School."

I hardly heard the rest. Radio School? There must be some mistake! Radio School? I was supposed to be a cook aboard a ship far at sea. I was stunned. The petty officer called out more names and announced where the others would go until he had gone through the entire company. "Cooks and Bakers School" was not mentioned at all. By the time he was finished, I was rethinking the designation. Radioman. That might not be bad after all. I'd still be aboard a ship, and in a radio room doing as important a task as feeding the crew.

I had no choice but to accept it. I had already quickly learned one didn't argue with the decisions of the Navy! I don't know what I figured a Radioman would be, but landing on the beach in that faraway dot of Iwo Jima was the furthest thing from my mind. I pictured the relative "safety" of the ship, but never did I imagine that a Navy radioman would be flat on his stomach in a shell hole on a volcanic island in the middle of the Pacific!

Each day, we were reminded why we were there, why the pushups and the physical fitness drills were necessary, why sighting down a rifle and making marksman could make the difference between survival or not. The need to obey orders unflinchingly was drummed home. In a time of crisis, a time of war, asking why or doubting one's actions could be the difference between life and death, not only for the one hesitating or questioning, but also for all the others in the group. Even the miniscule tasks like folding one's towel to specification were designed to test how one follows orders. The fate of the world could very well hinge on the fold of a towel.

The news from the front came on a daily basis, even though the battles we heard about were long over. The war definitely was no longer a game in the abstract. It was sobering to think that the men at sea in the midst of the battles had, themselves, been right where

I was, doing the same things for the same goal. Could I follow their footsteps?

An important part of being in the Navy was knowing how to swim. Swimming came easy to me, and I did not have a problem passing the swim test in the Olympic-sized pool. Jumping from the twenty-foot platform into the pool and swimming its length took no effort at all, as I had been on the swimming team in my high school. Passing the swim test was a requirement, and without passing it we were told we'd be denied our first liberty.

There was a friend and barracks mate of mine named Garner who wasn't so lucky. He flunked miserably, having never learned to swim more than a dog-paddle. Back at the barracks, Garner lamented, "I'm going to miss liberty, Geurin. This might be the last chance to see my mother before we're shipped off. What am I going to do?" He was a fellow Californian from Beaumont, about 85 miles northeast of Oceanside, an easy enough trip on a weekend pass. He was counting on going home; we all were. We were in the common area. It was our only free time when we got to write letters home, or read letters from home.

The mail was distributed once a day by a fellow from the barracks chosen by the petty officer. He got to run down to the main base post office, pick up the mail for our barracks, and distribute it.

My mail was eagerly waited for every day, not just by myself, but by the entire barracks, because when Mom wrote every day, she frequently sent cookies, too. I didn't get to eat many, if any, of them because as soon as the package was opened, I was suddenly very, very popular. I didn't mind; in fact I kind of liked the idea that my Mom's cookies were being eaten by all my barracks mates.

It was during one of those times I heard about V.E. There was nothing that could have prepared me for the news that came that terrible day. Those posted casualty lists I once read on the bulletin boards of the town hall back in Bakersfield took on an entirely different, and personal, meaning.

V.E., my cousin, my friend, who, with me, one day would open a bakery, was shot down over France and killed. The news was devastating! If there had been any lingering thought about war being chivalrous, glamorous or romantic, it was gone with that news. No idealistic visions of heroic swashbuckler then; just an image of my cousin going down with his plane over France. By then, Jimmy Smith, his younger brother, had also joined, a year ahead of me, and was in the Navy somewhere in the Pacific.

My mother wrote and told me what happened. Aunt Lulu had a nervous breakdown on hearing the news. She refused to believe that V.E. wasn't coming home. More bad news followed that first awful telegram. The smartly uniformed man who came to her door to tell her about the crash, also informed her how he had died. "Mrs. Smith," one of them said, and she looked past his shoulder, not wanting to look him in the eye, "your son was part of a brave crew. I know you are proud of him."

My Aunt Lulu didn't respond. She was still looking like she was not hearing what was being said. Finally she asked, "When can we bring him home?"

At that point, the three officers who were tasked to bring the sad news looked over her head at Uncle Vernon. It was then, Uncle Vernon told me later, that he knew the worse. "He and three others who went down with their planes couldn't be retrieved separately. There will be a ceremony in Chicago."

Aunt Lulu said nothing, and after a few more details, the officers left. My Uncle sat by my Aunt for several hours, trying to explain to her that when the airplane went down in a fireball, there was little left of the crew in it. Aunt Lulu kept shaking her head. Her beautiful son V.E. couldn't be gone, couldn't be a part of a burned heap of metal and ashes, fused together in a lonely box on its way back to the States.

When my Uncle got the news about where to go for the services for V.E. and the crew of that fateful flight over France, my Aunt had another nervous breakdown. She couldn't face it, couldn't face watching her son buried in a combined grave. Uncle Vernon took that sad trip alone to pay respects to his oldest son. It had to be the hardest trip he had ever taken, and it aged him prematurely.

Fearful that she would lose her only remaining child, his mother, my Aunt Lulu, pleaded with the Navy to release him. War becomes very personal when it's one of your own who doesn't come back. Contrary to popular belief, there was no "Sullivan Act" which allowed surviving sons to return home from conflict. Even though the year before, in November 1942, the five Sullivan brothers were killed when their ship was torpedoed during the Battle of Guadalcanal, and in spite of efforts to make it a law, there was never a law that prevented that from happening again. My Aunt's pleas were finally heard, and Jimmy Smith was released from service on a hardship discharge, even though that was not his option. If he was glad to be home, out of the fighting, he never showed it. There was that void that couldn't ever be filled, the loss of his brother. Nothing was ever the same again. A light went out when V.E. died.

It was one of those times when Garner and I were talking in the common room. He had just said, "What am I going to do?" in reference to his missing liberty and perhaps the last chance to see his mom in Beaumont before he was deployed.

I responded, "You know I could easily pass the swim test better than the whole bunch here." It wasn't above me to brag a little.

"Damn shame you can't take the test for me," Garner groused.

"Yeah," I said.

For a moment, neither of us said anything. I think we got the idea at the same time. Garner said, slowly, "Why not?"

"What?" I asked, not quite understanding what he had just suggested.

"Why not take the test for me? You could do it, Geurin. Hell, you and I are about the same size. Who would know?"

"You're joking, right?" But I could tell by his eyes he had a slimmer of hope, but would call it a joke if I turned him down.

"Geurin? How about it?"

I had just bragged I could out-swim the whole bunch here. How could I turn him down? It was chancy, and if I got caught, both Garner and I could face a dishonorable discharge. "Why not?" I answered with a lot more confidence than I felt. It seemed harmless enough, taking the test for him, and, at the same time, proving my proclamation was more than just bragging.

Garner pulled his dog tags over his head and handed them to me. They dangled between us on his fingers for a very long moment. There was still time to change my mind. I slipped my own dog tags over my head and handed them to Garner. I slid my finger under the metal chain of his dog-tags and put them around my neck. Before I could change my mind, I headed off to the pool again. Thankfully, with as many recruits going in and out of there every day, I wasn't remembered as having already taken the test. The only thing that counted was our performance in the pool, and passing the most basic of the requirements.

The sailor at the door checked the dog tags around my neck and checked off Garner's name on his clip board. He said, "Last chance, Garner." I nodded and went on in to the pool. My heart was beating fast, and I took several deep breaths to steady myself. So far, so good.

Remembering that Garner had had a hard time of it the first time, I deliberately fumbled, and tried to act as though swimming was foreign to me. I was instructed to jump in and swim across the pool. This I did, with flopping motions just enough to make it. After pulling myself up on the other side, I was rewarded with a nod from the observers. I—or rather Garner– had passed the swim test.

When I did that, I had no thought then that a sailor at sea aboard a ship that's severely damaged may have to swim for his life, and that the swim test was more than simply another chore to get through basic training. It was part of the life-saving techniques for which we were being trained. Doing the swim test for Garner was just a favor for a friend. I hadn't thought further than that, and neither did he. For Garner, it was just to ensure going home for liberty.

When I got back to the barracks and exchanged dog tags once more with Garner, I told him, "You passed."

He slapped me on the back, and a happier sailor there never was.

The first week was the toughest, getting acquainted with the Navy way of life and expectations. I never knew there could be so much marching, swimming, drilling, and attending classes. Classes. . . I thought I left that behind in high school, but, no, there was more studying there than I ever had in school. Every day, honor, courage, commitment were drilled into my head. They were words that had little meaning on the outside, but there they began to take on a deep and serious meaning.

The second week we hit the confidence course . . . it was also known as the obstacle course but, as our instructor explained, "The Navy doesn't let any obstacle stand in its way; these aren't obstacles. You will pass them. If you do not pass, you will do this course until you do."

Every part of the course was designed to simulate shipboard situations that we could encounter in an emergency. Passing the course was essential, not as just another task in basic training, but as something that could very easily save our lives later on when we got into the thick of battle.

The third week I boarded a training ship. Everything was hands-on. There I learned the ship nomenclature and first aid techniques, to semaphore (signaling with flags). This was the classroom lessons needed to survive in the real-life world of the Navy ships at sea. We were shown silhouettes of German and Japanese ships as well as our own. We had to learn them, and it was drummed into us how serious it was to tell one ship from another.

By the fourth week I was beginning to feel like I could do anything the Navy demanded of me. It was the first time we were with the Marines. We had to go over to the Marine base for the firing range. The Marine sergeant was not too friendly toward "swabbies."

He put his hands on his hips and shook his head. "Another bunch of swabbies," he said, including the Lieutenant and the Lieutenant Commander, both of whom had escorted us there. "Waste of my time." He growled, "You swabbies can't shoot straight, and you wouldn't know the business end of a rifle if your noses were stuck up the barrel." He looked us all over and spat, "The only real fighters in the war are the Marines." He said as he begrudgingly led us to our spots on the shooting range, "You swabbies ain't nothing but taxi service for the real Marines. You ought to stay in your place. Let the real fighters handle the guns." I was surprised that a Marine sergeant would have the audacity to talk that way, not only to us, but also to our Lieutenants.

In spite of the insults—or perhaps because of them – we were serious on the firing range by then. There was live fire, and targets both to hit and to avoid. There were instantaneous decisions made whether to fire on a target or not. "Friendly fire" became more than just abstract words. There were pop-ups of both the "enemy" and "civilians" with only a split second to decide on which, if either, target to take out.

By the end of the course, there wasn't a one of us "swabbies" who didn't pass the hard taught firing range. The Marine sergeant grudgingly, it seemed, signed the scorecard that gave us our marksmanship rating.

That week we also had our pictures taken. By then, my hair had grown out. I had a cowlick that wouldn't be tamed. When it was my turn, I pushed my cap back on my head and grinned into the camera. This was for Mom, and I wanted her to have a picture that would reflect what I felt just then . . . Navy all the way! That was another of those pivotal moments.

The following week things got more intense. We were near to finishing, and there was still a lot to learn. The war still went on, and what we heard was not encouraging. Bloody battles were being fought on land and sea. Ships went down and sailors died at sea. That type of news was not broadcast at the base, but I learned it from letters from home. What wasn't in my letters, was in others. We all shared letters and talked about what was waiting for us when we finished basic training. Instead of making me fearful, it made me all that more determined to finish basic and get on with it. My goal then was to get through tech school and into the war I had joined to fight.

Shipboard damage control . . . firefighting on board . . . how to escape smoke-filled compartments . . . all of this was piled on us that week. These skills were vital to a crew of a damaged ship at sea. I understood that my life, and the lives of all those on board, might well depend on how well I, and the others in my company, learned these techniques.

We were taken into a square, brick building that had nothing inside but cold, dark walls. There was the first time I put on a gas mask. I felt I was suffocating, but I kept it on and breathed deeply as the instructor pounded into us. I didn't know at the time if the haze rising up around me was real gas or not, and I didn't dare find out by removing the gas mask too soon.

That final week, we went over everything we'd learned so far. We took what amounted to our "finals." It was surprising to me how much I had learned. I was never much for studying the books. I was always the type who preferred physical exercise to classroom study. But there at the Naval Training Station it was different. This was more than the Three-R's. These lessons literally could mean the difference between life and death, not only for me, but also for my shipmates.

Boot Camp's main emphasis was changing young wet-behind-the-ear kids into sailors, and weeding out those who couldn't cut it. There weren't any in my group who were dismissed. I know at that time everyone there was as determined as I to make it, to be part of something that was far larger than our little group. The guys in my group came from all walks of life, but what we had been as "civvies" didn't matter much at Basic, and sure didn't matter a hill of beans when the real tests came in battle. Graduation from Boot Camp was just a stepping stone to something bigger and better, to the "real Navy."

There came the moment I will always remember, even if every other moment of Boot Camp were to fade away: Graduation. For the first time, I donned my dress blues. Was that me? That smart looking sailor in the mirror? They were all the same, these barrack companions, friends, mates who had gone through the grueling requirements of basic training with me. We were all transformed for that one magical time from scraggly civilians who couldn't walk a straight line to proud, tall sailors in dress blues.

The graduation ceremony was a blur. Navy band. Parade in front of officers and dignitaries. The knowledge that somewhere in the stands were my parents made me walk all the more proud to the parade grounds, and march straight and sure past the reviewing stand. It was a day of mixed emotions. I had made it! This part, this first crucial part, of becoming a real sailor on a ship at sea was over. Tech school had to be easy compared to what I had just finished!

Liberty! Every sailor's dream that he plots and plans for from the minute he enters the service! We were given three-day liberty right after boot camp, with orders to report back to our tech school. Since my tech school was on the same base as basic, and my parents lived up in Bakersfield, 240 miles north, I figured that was a breeze for me.

That first liberty was at home with my parents. I went with my parents after the graduation ceremonies. How strange it felt to be there. I had left there a young boy, fresh out of high school and returned a man. Home with Mom and Dad was at once comforting and oddly out of place. I felt I had outgrown being there, being in my own room. At the same time, it was sort of heart-wrenching to know I would soon leave there again, perhaps forever.

That first day, I slept in, even though I automatically awoke before dawn. My first thought was to jump out of bed and dash out to do my calisthenics! It took all of five minutes for me to realize I was in my old bed in Bakersfield, and rolled over and stayed in bed.

Those three days passed all-too-swiftly, and it was back to San Diego. Dad drove me, and when he left, I felt an upsurge of home-sickness for the first time since I'd enter Basic almost two months previously.

CHAPTER 7
Radio School

Graduation from Boot Camp was just a stepping stone to something bigger and better, to the "real Navy." I returned to the Naval
Training School in San Diego, and reported to Radio School on the
opposite side of the base from Basic training. Now was knuckle-
down time. When I was home, I heard first hand what was happening on the other side of the world. My particular interest was in the
Naval battles, and they were being fought fast and furious.

The battle of Tarawa in the Gilbert Islands, 2,500 miles south-
west of Hawaii, was being hard fought in November of 1943. It was
hard fought and resulted in high casualties for the Marines, and a
higher cost to the Japanese. Before the battle was over, the Marines
suffered nearly 3,000 casualties . . . a number unimaginable to me.
The Navy had taken the Marines there and participated in the shelling of the islands. The fierce battle was started when the big guns
on the USS *Mount Olympus* (AGC8) (amphibious force command
ship) boomed across the island. I heard all about that terrific battle
while I was home, and knowing that the Navy played a pivotal role
in the battle only inspired me more and reinforced my decision to
join the Navy.

The thought of being a cook aboard a ship at sea still appealed
to me. I was already assigned a specialty—that of a radioman. While
it was intriguing in some way, I knew my strength lay in baking, and,
after all, that was the goal I had set for after the war. I figured noth-

ing ventured; nothing gained. I figured that if I asked for a reassignment prior to going to the first day of Radio School, I would have a better chance. After all, Cooks and Bakers School was promised to me, I argued with myself. Shouldn't I at least have a chance to get it? I convinced myself that I could present a compelling argument.

At the Radio School, I approached the Ensign and asked if I could have a word with him. Then, before I lost my courage, I said, "I'd like a transfer to Cooks and Bakers School, and be assigned sea duty. I'd liked to be on board a Navy ship—a destroyer, cruiser or any other Navy vessel. I'm not shirking duty," I said. "I just feel I'd serve the Navy better as a cook."

"Geurin," the Ensign said, "if you drop out of Radio School now you will be sent to Oceanside."

Being sent to Oceanside didn't mean much to me. I said, "What's that?" From his tone I didn't figure Cooks and Bakers School was in Oceanside.

"Geurin," he said, and I swear there was a wicked look in his eye, "Oceanside is where the amphibious forces are. Marines, Sailor. You will be trained to hit the beach with the Marines instead of staying aboard a ship. That what you want?"

The prospect of being a radioman suddenly became a little more enticing. "No, sir!" I answered. I pictured myself staying in the radio cabin, taking and sending ship-to-shore messages. It was a much better prospect than going ashore with the Marines. I knew "hit the beach" did not mean sunny beaches and beach balls and admiring bathing suit clad beauties!

For three months, I studied and paid attention, all thoughts of Cooks and Bakers School finally put aside. The radios back then were not miniatures, but huge affairs that could take up an entire room. The portable radios were not pocket-sized, but large, unwieldly three-piece components. Each piece weighed approximately 40 pounds. Three of us trained together to assemble the radio, quickly and accurately. Even though all of us were trained on all the radios, I concentrated mostly on the ship radio. After all, that was going to be my station . . . in the radio room, decoding messages from shore-to-ship or ship-to-ship. The large radio was at the front of the room, and I sat in one of the desks with my earphones clapped to my head and listened and decoded, and tapped and learned.

What a difference between basic and tech school! We had almost unlimited freedom, as long as we were in the classroom on time and stayed for the lesson. Evenings and weekends were our own. For young men out on their own, most for the first time, this was the euphoria of our lives. In spite of having that freedom, what we didn't have was the right to leave the base whenever we pleased . . . unless we had a pass. Ah, that was the rub.

There was a fellow on base who had a profitable side business. He sold passes. How he did it, I never questioned, but for $25 I purchased a pass that allowed me off base any time I wanted, weekday nights or weekends. It was the best $25 I ever spent!

With that pass, the world was my oyster. Or at least, California was. A little over 114 miles north of San Diego was a sailor's paradise: Long Beach, California. It wasn't just that it was a Navy port town with big ships out in the harbor that drew sailors, but also their unique amusement park called The Pike. It was right on the ocean, and had a roller coaster which swung out over the water. A Navy-friendly place, it offered a fantasy respite from the war. Sailors walking arm-in-arm with their young ladies was a commonplace sight.

The Pike had everything, from roller coaster and Ferris wheel, to hawkers daring passersby to "Knock down the dolls!" or "Shoot the target! Got a good eye, Sailor?" There was the popcorn popping in festive colored, brightly lit popcorn machines. There was the loud music and clanging bells and overlapping chattering. It was a place where a sailor could lose himself for an hour or an evening, and forget for a while that there was a war over that same ocean.

To get to Long Beach from San Diego all I had to do was hitchhike. Hitchhiking in those days was easy for anyone in uniform. There was nothing safer, or faster, than putting out your thumb on a busy highway while dressed in a uniform of the United States Armed Services. Drivers of all ages, and in all kinds of vehicles, were eager to give a ride to a serviceman. It was their way of doing their part, or thanking us for our service. Many of the drivers had someone in service and could empathize with us. Some days, while walking backwards down the highway with my thumb out, I'd see agriculture trucks crammed, not with cabbages from the field, but cheering and shouting soldiers and sailors alike.

The return trip was just as easy, only then I was more likely to see sleeping sailors in the beds of commercial trucks, heads rest-ing against crates of eggs or noisy, caged chickens on their way to market.

Back at Radio School, one of the major requirements was to be able to type at least 40 words per minute. That was a struggle for me, for I was never one much good at a keyboard. I preferred boxing and swimming to staying inside and typing. I never took typing in high school. In my day, typing was for the girls who planned secretarial careers. I didn't know any of my male friends who took typing in school. Taking it now wasn't something that particularly set well with me. I felt I was all thumbs. Another major requirement was learning Morse Code. I was like every other kid when I was young and played at *dot-dot-dash-dot-dot* without ever knowing what it meant. Learning that Code was much harder for me than my hunt-and-peck typing method. But I preserved, and in spite of my self-doubts, one day it all became amazingly clear. It was like suddenly a cloudy window was opened and I could clearly see the answers. It's always like that when something finally clicks . . . I wondered, *now why was that so hard?*

The most difficult requirement at Radio School was learning the specialized mathematics of Ohm's Law. Never one who enjoyed any form of mathematics at school, this was a definite challenge for me. The understanding of Ohm's Law was essential in the operation of the radios. The most simplistic definition, which I was encouraged to learn, was that the potential difference across an ideal conductor is proportional to the current through it. Once this central basic point was pounded into my head, the principle of Ohm's Law became, if not clear, at least understandable enough to get me through that session of Radio School.

At the end of the three months of compressed, accelerated and intense training on the radio, there was no big celebration, no parade with friends and family cheering us on. My final test was a hands-on timed procedure. I had to decode an incoming message, as well as send a coded message to another ship. The room was crowded and the pace was fast, just as it would be aboard a ship.

One of the amazing turn of events was that all of us, myself included, not only learned what would have been foreign to us just less than a year ago, but learned it well enough to become Radiomen in the United States Navy, responsible for vital communications during a world war. We entered the Navy as boys, seventeen to nineteen years old, with vague, idealistic views of what it meant to be a sailor. We were now men on the verge of becoming an essential part of fighting, and winning, the war. There was no better feeling than to realize that I was an important part of that.

That last day of Radio School, I, along with the rest of the 60-men compliment, eagerly awaited learning where next we would go. I imagined being on a ship at sea, having more hands-on training in the radio shack, while the ship eased out of port and headed toward the war zone. Which ship and which theatre of war, I did not know, and could only anticipate, but in the long run, it did not really matter. It would be somewhere important, and the time at the Radio School would prove to be something I could put to good use. The war had not eased during the time I spent in Basic and Radio School, making me only that much more anxious to become a part of it and a part of the final solution.

Finally, the day came when I graduated, and breathed a sigh of relief. Now, now a ship at sea! Now, all of what I endured the months at Basic and then at the Radio School would be put to good use. I felt great! Ready to take on whatever the Navy had planned. There was nothing that was beyond me now. I was a Radioman in the US Navy!

That euphoria was short-lived. That day, just as I was inwardly planning my next move, the chief petty officer who had been our primary instructor came into the classroom and said, "Listen up!"

Chairs scraped back, talk quieted, and we sat waiting for the pronouncement of where we would go. I knew that the class would break up, and we would be assigned different ships, but that was all right. Friendships made in Radio School would stand the test of time, I was sure.

"Beginning tomorrow," the chief petty officer announced, "this class will meet at 0800 in Oceanside at the Amphibious Forces Training School. You will train with the Marines. That is all." He turned around and left us stunned. So much for getting out of hitting the beach with the Marines! I inwardly groaned, thinking of the long ago promise of Cooks and Bakers School.

CHAPTER 8
Amphibious Training School

The following morning at Oceanside, I reported to the Amphibious Training School. My group was not the only Radiomen sent there, and the entire combined group was divided up alphabetically, three radio operators to a beach party. The beach party was more than just the radio operators, and we all had to learn to work as a team. There were forty-six men to a beach party, which included radiomen, signalmen, boat handlers, and the coxswain (who steered the boat ashore). The goal was to become one unit, flowing as seamlessly together as the waves that washed upon the shore of that faraway island.

On hand to greet us were the two men who would be our commanding officers, not only throughout our training, but also throughout our engagement in the war. Lt. Commander Ernest duPont, of Wilmington, Delaware, and 1st Lt. Frank Skubitz, of Ely, Minnesota.

The main objective of our training was to become coordinated with the Marines, who were in our boats every time. The idea was to transport them, along with our beach party, to the beach, and be able to communicate smoothly between the two services. There has always been a rivalry ... sometimes not too friendly ... between the Marines and the Navy. This type of training was essential to the eventual success of taking, and holding islands out in the Pacific, and we had to put our rivalry aside. Even though we practiced with them every day, we sailors still kept pretty much to ourselves when it came to our free time.

One of the few good things about the way we were assigned was that Garner was assigned to my group. Our third radioman was Leon Estrada from Fabens, Texas. The radios, modern for the times, were big, clumsy and took all three of us to put together and operate. The importance of quickly setting up the radio, running up its antenna, and making contact with the ships was stressed. We would be the main contact between shore and ships. It wasn't overlooked, the fact that radio antennas were beacons for the enemy's version of radar, but that little fact was underplayed. After all, we were United States Navy!

Our evenings and weekends were free to study or raise hell. Few of us chose to stay in Oceanside every evening with our nose to the grindstone! It probably would have been better all around, but we were still young enough and brash enough to think we could carouse in the evenings and still be fresh enough and bright minded enough to learn what was needed the next day in class.

On one foggy weekend morning, I borrowed a car to go into San Diego and the USO. Cars were few and far between, but I had a friend who was on duty that weekend and was willing to lend me his car, for a small price. The cars had running boards and big, wide fenders, and huge bumpers in the front and back. I had my pass handy when I reached the gate, and a half dozen of my friends hanging onto the back bumper and running boards. The guys on the running boards jumped off just before we reached the gate and casually walked through when the guard's attention was diverted when he approached the car. I jauntily showed my pass, and the guard nodded me through. Just as I started the car again to ease through the gate, I had to stop for a train. The railroad tracks ran parallel to the base, and the crossing was right across from the gate. The clanging of the warning signal sounded like the voice of judgment. The guard started to go back to the guard shack, then made an abrupt stop and looked at the back of my car.

I heard the guard shout above the blasting of the train's horn. "HEY! HEY YOU!" The two guys who had been clinging to the back bumper for a quick, unauthorized trip into town were caught! The guard ordered them. "Off the bumper! Show your passes!"

They didn't have them.

"Back to the base! Now!" The guard ordered.

I sat there, hands frozen on the steering wheel, feeling sweat edging under my collar, sure the guard was going to reprimand me for aiding in sneaking them out of the base. Visions of spending the weekend in the brig, instead of dancing with the women at the USO, washed over me.

When the train passed and the guard still did not come after me, I breathed a sigh of relief, and gunned the car across the tracks. My other friends, who had jumped from the running board, were waiting on the other side of the tracks. They were laughing, while saying, "Come on, Geurin! Hurry before he gets wise."

Luckily for me, the fellows who had been caught never told the guard that I knew they were back there.

Our training was far more realistic than any we had so far. We went aboard the amtracks, which circled out into the ocean until they were in a line. Then they went in together as a wave, just as they would in a real battle. At a certain point off shore we, and the Marines, went over the side of the amtrack into the water. We were in full gear, our radio parts protected in a waterproof webbing. This was as close to the real thing as I had gotten. It gave me a real taste of what would be expected "over there." The coxswain guided the now-crowded craft on the wave toward the shore.

Not far from shore the instructor, a Marine in full gear, shouted, "Hit the beach!" I scrambled over the sides of the lurching craft and splashed to the beach. We three Radiomen gathered quickly together and dug a foxhole. After we figured the hole was deep enough, we started assembling the radio as we were taught. It didn't always go smoothly, but after the first try, we always managed to get the huge radio together. Just as quickly, we sent our position back to the ship, alerted them what was happening on our part of the beach.

The call to "Back to the boats," and we tore down our radio, filled the foxhole and retreated to the landing craft upon command from the instructor on the beach. We returned to the amtracks, only to go back to the beach, again and again, all day long. It never quite became routine, but it did become easier with every repetition.

It was a warm, fall day in October 1944, after six months training at Oceanside, hitting the beach every day in mock invasions, setting up our radios, living in Quonset huts, that we were told we had passed all the requirements and were ready. There was little time for celebration. No pass this time. What we did get was our rank. I became 3rd Class Radioman (RM/3C). The lot of us were given just enough time to gather our gear into our sea bags and report to the gate for departure.

CHAPTER 9
USS *NAPA*

The large, white Navy buses pulled up to the gate and the hiss of their brakes as they stopped sent conflicting emotions through me. Here I was, all finished with stateside training and ready to go wherever the Navy would send me. *Where would that be? What if I had to stay stateside?* I was no longer naïve, and knew that, in spite of training to "hit the beach," that did not necessarily mean I would be immediately sent overseas. I could be with some reserve force, waiting for the call to be a replacement. I knew nothing more than our beach party was certified ready to be assigned to a ship.

Sea bags were stored in the cargo cavity of the bus, and I climbed on board. It reminded me of how I had gotten there in the first place, only this time out I was in uniform and a fully trained Radioman. While I knew on that first trip where I was going, I didn't know this time. None of us did. That we were heading away from training into the real war was enough to know.

As the bus droned northwards, in spite of the anticipation and uncertainty, I fell asleep. A little over two hours and 85 miles later, I awoke with a jerk when the bus slowed and pulled into the still-brand-new Union Train Station in downtown Los Angeles. It was built in 1939, by the unusual cooperation of the three major railroads, the Atchison, Topeka and Santa Fe Railways. The Union Station was the showcase of train depots nationwide. Outside, the clock tower was the centerpiece and could be seen from miles around. There were no high-rises in Los Angeles to obstruct the view. Inside, it gleamed from highly polished marble floors. Our march-in-time footfalls echoed as we marched across the floor, following the Chief Petty Officer to the tracks.

There, a puffing black steam engine sat hissing on the tracks. It sounded as though it were anxious to be out of there. We formed one line and got aboard the train. It was a troop train, and no one but uniformed personnel were aboard. We were still not made aware of our destination.

I stored my sea bag overhead and stretched out in a window seat. Garner was able to cram into the seat beside me. For the first time since finishing amphibious training in Oceanside, I relaxed. I didn't particularly care where the train was going. The important fact was that I was, at last, on my way to the "real Navy."

"Garner, this is it! On the way to the real thing."

Garner grinned. "Yeah, know what you mean, Geurin. Seems it took forever, but here we are." He nodded toward the window. "God knows where!"

The train rumbled and chugged its way out of Union Station. I relaxed, because at that point it didn't matter where we were going. As the train wound its way northwards, the territory began to take on an achingly familiar look. Over 110 miles later, we pulled into Bakersfield. Nowhere near as grand as Union Station in Los Angeles, the Bakersfield station was a one-storied, near-the-track station fairly standard of railway depots of that era. Unfortunately, there was no time to get off and touch familiar ground. I stared from the window as additional troops came aboard. I could imagine Mom and Dad out there, in the house in back of the bakery, thinking about me and not having the vaguest idea I was in the same town, on a train stopped at the train depot. It was all-too-soon that the steam engine blew its whistle and started northwards once more.

The train trudged onward. There were some who read books; others played cards; still others took the opportunity to sleep. Small groups formed to exchange the latest news from the war front or from home. I played some cards; slept some; and awoke with leg cramps. My long legs weren't meant for the squished space between train passenger seats! There was nothing comfortable about the troop train.

Meals were plain and served in the galley. I picked up a sectioned metal tray, walked down what amounted to a buffet line, and squeezed through a narrow aisle to eat in the dining car. It wasn't bad; much better than the chow hall at Basic.

Sleeping was where and when we could. I mostly stretched out in one seat, with the back of the seat practically in the lap of the person behind me. He, in turn, had his seat back; and so it went until the hapless person in the far back seat had nowhere to go but to put up with the seat in front of him practically in his lap.

On that seemingly endless train ride northward, the weather changed and the unheated train became about as warm as the inside of a refrigerator. I discovered if I took out a couple sea bags from the overhead storage compartment, I could squeeze in there and it was much warmer than curled in an impossible ball in the too-short seat.

The train climbed through a mountainous pass, its whistle sounding lonely and ominous. There was snow out there; deep piles of snow clinging to the mountainsides. I was certain that the rumble of the train and the echoing of its whistle would bring the entire mountainside down onto the train, or, at least, the tracks. I stared out the window at the never-changing landscape . . . miles and miles of nothing but snow clad mountains and trees bent under the weight of ice. The black smoke from the train blew back and gave the scene an eerily black-and-white portrait feeling. By now, many of us had dug out our peacoats against the bitter cold. The only defense we had against the chill was to wrap gloved hands around hot cups of coffee. The galley kept the coffee pot going. The coffee tasted like sludge after a while; I'd a suspicion that the same grounds were used over and over. But it was hot and that was the priority, not its taste.

That trip seemed endless, but finally, after a day and a half, the train started slowing. The hiss of its brakes as it wound down the mountain and into civilization again brought a fury of activity. Seats were sprung back into sitting positions. Sea bags were dragged from the overhead compartments, or removed from under heads that had used them for pillows. Uniforms tucked and tugged to some sort of semblance of neatness . . . not that any could meet the uniform neatness standard by then. My sea bag was on the floor, squashed between the seats, in order to make room for an impromptu bunk in the overhead cargo space.

Those of us by window seats looked out at the signs of a town coming into view. The others stood in the aisle and peered over. There soon came the white board sign on a black pole that announced "Astoria."

Astoria, Oregon, was one of the oldest settlements west of the Mississippi. It had been a military port since 1805 and the Lewis & Clark Expedition. Located on the Columbia River in the northwest coast of Oregon, the port was thirteen sea-miles from the Pacific and was home to the launching of many of the Pacific Fleet ships.

When the train settled, the Chief Petty Officer ordered us to get off. I picked up my sea bag and joined the line in the crowded aisle and worked my way to one of the exits. There, I jumped to the train platform, fell in with my company, dropped the sea bag at my feet, and waited, and waited, until everyone had disembarked. By then, I had figured out "hurry up and wait" were the keywords to success in the Navy!

From the nondescript train station in Astoria, we climbed aboard yet another Navy bus. Sea bags once again in the cargo hold, we were separated into companies and ordered into specific buses this time. Somewhere out in the harbor was our ship! That much, we were told. The men aboard this bus would all be on the same ship. That was exciting news! Garner and Estrada would be on the same ship as I. It was coming closer and closer, this triumphant end to training and the exciting beginning of being onboard an actual war ship.

As the bus trudged along the city streets toward the port, I couldn't help but wonder what ship and what theatre of war. Unknown to us, the ship waiting there was built in Portland, Oregon, by the Oregon Shipbuilding Corporation and ours was to be its first crew.

Our ship began to be built in June 1944 and was ready for sea by August that same year. That was a fast turnaround from a roll of blueprints to a fighting ship, but it was time of war and the ships were needed fast. On 12 August, the metal plates binding the ship to the port were severed, and the brand new ship slipped down into the sea after being Christened by Mrs. Cranston Williams, wife of the General Manager of the American Newspaper Publishers Association. From there, it went to a dock to be fitted for her role in the war.

On 27 September 1944, the newly Christened ship made her first test run down the Columbia River from Portland. She had aboard her a small part of the men who would be manning her, taking her into battle. On 30 September, she cruised down the river to Astoria, where she patiently awaited the coming onboard of the rest of her crew. She was now an official part of the United States Navy.

Later, I found it apropos that the USS *NAPA* was being built during the same period I was in training to eventually board her. It seemed like fate!

When I first saw the ship, I figured it had to be a good omen. It was the USS *NAPA*, APA 157! What better sign could there be for a California boy than to be aboard a ship named after one of the most prestigious areas in California—Napa Valley! There were forty-six men to a beach party, who were part of the Ship's Company, a 400-man complement. It was exciting to know that we would be part of the first crew to man the ship. Even before we saw her, she became "our ship," and, soon, "my ship." I didn't even begin to wonder what journeys I would take aboard her, but I knew it would be nothing less than exciting and rewarding.

It was 1 October 1944, and officers and crew, including the Beach Party, assembled at quarters on the after boat deck before invited guests, relatives and friends. The sound "NAPA NAPA" resounded over the public address system as, for the first time, Captain A.H. Ponto, United States Navy Commanding Officer for the Naval Station in Astoria, came aboard for the commissioning ceremonies.

The ceremony was impressive and well choreographed. The band played the National Anthem and we all snapped proudly to attention. After a prayer by Chaplain Malcolm Eckel, which ended with, "God bless the crew of this ship, and God bless America," commissioning orders and remarks were made by Captain A.H. Ponto, who had seen the ship from Portland to Astoria. The ship was formally handed over to its first captain, Captain Francis J. Firth, from Long Beach, California.

"The Captain extends a hearty welcome to the officers and men assigned to duty in the USS *NAPA* and hopes that our duty together will be happy and fruitful. All of you realize, no doubt, that our missions will be very essential and most important. There will be no easy jobs nor luxury cruises. Therefore, there is no place on board for weaklings or shirkers. To accomplish those missions, sincere thought and consideration must be given to the principles of our ship. Put them into practice and you have an APA of which both you and the Navy can be proud." So said our new Captain. In all, there were 446 members of the crew.

The ceremony itself was, to me, the final step in my becoming a United States Sailor! Here I was, a Radioman in the US Navy about to sail out to sea in a brand new ship, and not just any ship, but the USS *NAPA*! I don't remember the entire ceremony, focused as I was on this new reality. In miles, I wasn't too far from Bakersfield or Visalia; emotionally and physically, I was a far different man than the seventeen year old boy who bugged his Dad to let him join the service!

The day after the official ceremony, working parties assembled on the dock and begin loading the first supplies aboard the USS *NAPA*.

By 6 October 1944, I was becoming familiar with the ship, and especially the radio room. The radio room was nothing fancy. It was a small, crowded rectangular room, crocked full of radio equipment. There were two straight-backed metal chairs and the rest of the area was occupied by the transmitters and receivers. There was barely room for two guys to pass behind the chairs. The radio itself took up the whole room. The only way in was through a small hatchway. At a little over six-feet in height, I had to duck each time I entered or left.

We still had not moved out of Astoria, but the time there was being put to good use as the USS *NAPA* was being readied for its time at sea. That day was the first payday aboard the ship. When the Boswain's pipe sounded, we all felt like we had arrived at last!

Nights aboard the ship took some getting used to. The bunks were two decks down. They were five high, stored against the sides of the ship. Each bunk was fastened to the ship and folded up against the bulkhead into a "V" shape when not in use. There were bunks on either side of the narrow aisle way. I quickly chose an upper bunk. They didn't look too stable and I sure didn't want someone tumbling through one, or off one, and onto me! During off-duty times, I soon discovered that the crease of the "V" of the folded bunk was an excellent place to catch a little extra shuteye without being caught.

By 12 October, the USS *NAPA* was ready for its first official cruise. Out of the Columbia River into the ocean for the first time, the structural test-firing of the guns began. The big **BOOM!** of the guns rattled the entire ship, but proved they were in working condition and well secured to the ship. By midnight, the wailing of a foghorn added another a lonesome touch.

None of us knew where this ship would take us, and none of us could have possibly even imagined that would be that mound of dark volcanic ash so close to Japan. Even the name of that island was unknown to us in October 1944 as we stood on deck on the USS *NAPA* APA 157 and prepared for the shakedown cruise. It was, for me, a good omen that the ship to which I was assigned, this brand new shining ship, was named after Napa Valley, California, so close to where I grew up.

We sailed north to Seattle, Washington, on that shakedown cruise. By 16 October, the USS *NAPA* was moored at Pier 91 in Seattle, Washington. It was then that the USS *NAPA* had its first crew injury. In looking back, it was, in a way, a good practice for the crew, although, to be certain, not pleasant for either S/2C Benjamin Ramsey from Douglas, Arizona, or S/2C Joseph Sandovich, originally from Toole Utah. Both of them were injured by winches. The medical corpsman of "H" Division got hands-on experience, which would come into play much later when we were at sea.

In Seattle, the USS *NAPA* took on much-needed supplies and even more needed ammunition. The three days in Seattle were spent all aboard ship. Somehow, I didn't miss going to shore for I was much too busy learning my tasks in the radio room. Being in an actual radio room onboard ship was far more exciting, and far more exacting, than even the practice on the ships in Oceanside. This was the real thing, and it didn't take long to realize what I did here could have far-flung and lasting effects, on not only the ship, but also the troops on shore.

On 19 October, with growing confidence in the ability of the ship, from Seattle we went south to San Francisco, California. There was an eight-hour layover in San Francisco, but even before we got there, the announcement of "NO LIBERTY" came over the loud speakers. As we sailed under the magnificent Golden Gate Bridge into the Bay of San Francisco, seeing that grand old city and knowing that we couldn't go on liberty was hard to take.

The main purpose of being in San Francisco was for the Boat Group to take on their Higgins and Cris-Craft assault boats. The Higgins were the amphibious landing boats, including the LCTs, LCPLs and LCMs. They were built by a small boat company out of New Orleans, Louisiana, owned by Andrew Higgins. He at first had a difficult time selling his idea to the military, but soon the name "Higgins" was synonymous with the landing craft.

When Higgins started his company, he employed less than 75 workers. By 1943, he had seven plants and employed more than 25,000 workers who were busily engaged in building the finest attack boats of the war. Higgins also had the distinction of having not only one of the first racially integrated workforces, but also from groups not generally thought employable: blacks, women, seniors, people with disabilities, and poor whites. Not only did he employ a unique workforce, but he also paid equal wages to all, according to their job function. It didn't matter if a worker was white, black, or female, senior or had a disability. If he or she did the job, he or she got the same pay. His faith in this multi-ethnic crew proved well deserved. By the end of the war, Higgins had turned out more than 20,000 boats, 12,500 of them LCVP (Landing Craft Vehicle, Personnel) attack craft.

The Cris-Crafts loaded that day were also LCVPs, and were built by the same company which would later be so widely known for their luxury and sports watercrafts. Back then, as with most of the major American industries, their manufacturing lines were focused on the war effort, instead of civilian water craft.

While we were at port in San Francisco, the Battle of Leyte Gulf was being fought in the Philippines. The Navy played an important and pivotal role. The amphibious assault on Leyte took place on 20 October that year, and by 26 October it was clear that the Japanese Naval forces would be defeated. While we cheered, it was, nevertheless disappointing because the Japanese Navy was all but dismantled. They had taken up the challenge of the Ally Forces, and the Japanese Combined Fleet converged on Leyte Gulf to fight the Americans and maintain control of the Philippines. Although the Japanese Navy fought hard and, at times it seemed they would win, they were defeated. The Battle Leyte would go on for two more months of fierce ground battle before the Allies declared victory.

The war was going on without us! We had actually looked forward to taking on the Japanese Navy, and now it didn't look like we would get that chance. It's funny, looking back, at how that prospect affected us. We were not sorry that the Japanese were being taken down, but at the same time, we wanted a piece of the action. It seemed, for a time, that all of our training would be for mop-up. How idealistic we were then!

After eight hours in San Francisco, when all the landing craft were aboard, the USS *NAPA* set sail again and headed south. Two days later on 21 October, the USS *NAPA* anchored in San Pedro Bay, California, over 400 miles south of San Francisco.

At San Pedro, the shakedown continued. This time there were gunnery exercises, tactical maneuvers, loading and unloading of supplies and men, debarkation and assault landing drills. Captain Firth worked the crew hard, and at times there was grousing. But we knew that the tough drilling was essential to make sure that the USS *NAPA* and her crew were ready for duty.

In spite of the rigorous work schedule, there was liberty at San Pedro. In fact, some of us managed to hitchhike into Long Beach, Los Angeles and even Hollywood. It was only a little over 7 miles to Long Beach, so that was the most popular of the places to go. The 31 miles to Hollywood was an easy trek because men in uniform never lacked for a ride, either to or from their ships. I didn't risk hitchhiking further than Hollywood, although I would have dearly loved to have gone to Bakersfield. At 135 miles northeast of San Pedro, it was too far to risk while we were on active training. I chose, instead, to go to Long Beach or into San Diego to the USO.

Maneuvers continued and intensified. On 23 October 1944, the USS *NAPA* took on its first fueling since it had left the berth in Astoria, Oregon. Somehow that fueling operation, in which almost 210,000 gallons of fuel were received, gave me renewed hope that soon . . . soon . . we would finally be on our way to the war. I was getting itchy by then to get on with it. Enough training, already! It seemed I had been in some sort of training my whole Navy career, and I wanted to get out there and do something a little more meaningful than the practice runs! Under all of that impatience, I knew the training was necessary. I didn't think so far ahead as to realize that it may one day save lives, my own included.

Nineteen days into our training, on 11 November 1944, the USS *NAPA* moored at Pier C in San Pedro for what was to be her final overhaul and inspection. The electricians from Divisions "E" and "M" were hard at work, checking and rechecking every part of the ship. It was hard to believe that many of the electricians making sure everything was shipshape had been wet-behind-the-ear, inexperienced, teenaged fresh-out-of-high school boys not so very long ago. Like the rest of us, they grew up fast, and learned their skills at an amazing rate.

I still didn't know where, or when, we would actually get into international waters and on into the war. It seemed everyday there was something else to add, or repair, or to practice. Meanwhile, the war was going on without us. We had a ship; we had the tools to fight; now we had to do was get on with it!

Finally! On 20 November 1944, the USS *NAPA* pulled up anchor from Port San Pedro and began a northward journey. Approximately 85 miles later, we docked again, still in California, at Port Hueneme. We picked up ammunition and the 30th Naval Construction Battalion, also known as the Seabees. These guys set the discipline and precedence upon which subsequent units were judged. That crew was rumored to be capable of building bridges out of toothpicks and mud that would withstand the weight of heavy tanks. After watching them, I began to believe the rumors were true. They were an older cocky bunch, sure of themselves and their abilities, and they had every right to be. They came from construction and engineering jobs straight into the Navy. They were skilled workers with the creativity and imagination it took to build something out of nothing. They all were rated from 3rd Class Petty Officers to Chiefs and commissioned officers.

CHAPTER 10
Pearl Harbor

Five days later, we got orders to leave Port Hueneme for Hawaii. At last, we were on our way to our first step into war—Pearl Harbor, Hawaii, where, on 7 December 1941, less than three years before, the Japanese had flung us into the war by bombing our base there. 25 November 1944, we departed Port Hueneme en route to Pearl Harbor. War games were over. We were in this for real, and the approach to Pearl Harbor hammered that home as nothing else could.

Up until the time I actually boarded the USS *NAPA*, I really had no idea of what war was really all about. I didn't know why Japan had bombed Pearl Harbor or what the actual consequences were. I was an idealist, viewing war from afar, knowing only that I had to be a part of it. I didn't realize how deadly serious it was, and how personal it would become. No matter how many accounts I heard of the war, no matter how many casualty lists I read, there was still that impersonal feeling that these were just names and numbers, not real people who fought, and bled and died. That my name could possibly be added to that list some day never even entered my mind. At nineteen, I was invincible.

On 2 December 1944, as we came into the harbor at Pearl, we passed the sunken and overturned battleship USS *UTAH*. We all stood on deck as the USS *NAPA* made its way into the harbor. It was my first glimpse of the treachery and aggression of the Japanese attack of 1941. Three years later almost to the day, and the scars were still achingly visible. There were other grim reminders of what had happened, and why we were there. The reasons for all the months of training became crystal clear. The war was very much on our minds. There was an overall feeling of sadness for what had happened there.

The crew was unnaturally quiet as we passed the watery grave of the USS *ARIZONA* which still rested on the bottom, part of its crew of 1,100 men entombed. It reinforced us, and made me very aware of our mission. For me, as I stood there looking out over the upturned American warships, I could almost hear the whine of the planes and the burst of the bombs over Pearl Harbor. Even then, the USS *ARIZONA* became a symbol of the reason for the war, the reason we, the United States, must win. I had no doubt that we would be victorious. I did not then realize the cost of the victory. The cost here on Pearl alone had been enormous. I knew that nothing less than victory would be acceptable.

A very long time ago, it seemed, I had thought about going to Hawaii, to the beaches of Maui, where dancing girls in grass skirts would greet me and smile as they put leis over my head. I never imagined my first steps on Maui would be, not to the tune of Hawaiian ukulele music, but to the deep no-nonsense voice of a beach party training commander. No dancing on the beach, no luaus. In truth, they were the furthest things from my mind at that time.

It was time to get serious. Every day, we made practice invasions with the Marines on the beach of Maui. We set up and took down our radios on the beaches in preparation for the real thing. We had no idea of where that might be, except that it would be somewhere in the Pacific. I envisioned an island very much like Maui, with golden sands and palm trees. After all, what was the point of practicing on that type of beach if it were not similar to our final destination?

On 12 December, our Captain Firth collapsed on the bridge of the USS *NAPA*. I was with the beach party on another one of our practices when it happened. Upon return to the ship, I was to learn our captain had abdominal pains, possibly appendicitis, and had been transferred to the hospital at shore. Two days later, we were told that our Captain would not be returning.

Commander Guido F. Forster, originally from Summit, New Jersey, was selected by the Commander Service Force of the Pacific Fleet to take over as commander of the USS *NAPA*.

The day before Christmas in 1944, we were set ashore on an uninhabited island just off Maui. We were there through the holiday. It was the loneliest Christmas I had ever spent. I did not even have the pleasure of spending it on board our own ship with our crew. The isolation, although not appreciated at the time, was one more way of molding us, preparing us, for the invasions to come. It was definitely not a Hawaiian vacation or a Hawaiian holiday! Our beach party of 46 included the radiomen like myself, signalmen, boat handlers and medics. None of us expected what would happen at Iwo Jima. The practicing at Hawaii couldn't prepare anyone for the black sands of Iwo Jima.

On Christmas Day, the USS *NAPA* was underway once more on practice maneuvers to an uninhabited beach in the Hawaiian Islands. The lowering of boats to simulate the debarking of troops was practiced over and over again until it became almost routine. We also "made smoke," a technique to hide our ocean footprint and confuse the enemy by concealing our ship in a blanket of smoke. Finally, the landing crafts were sent back to the ship and hoisted aboard. That is ... everyone but the beach party. We spent that night on the island, setting up and sending messages back to the ship.

27 December 1944, elements of the Fourth Marine Division were loaded and the USS *NAPA* got underway for maneuvers with the Marines again at Maui. We practiced for days there. It was difficult to think of the actions we were taking as a prelude to an invasion force on some other faraway island. It was just hard work, and I would rather have been on liberty than hitting the beach in Maui with the Marines. Ironically, the more we practiced, the more remote the war became. The feeling I had when we pulled into Pearl Harbor and saw the ravages of the attack of three years previous also had slowly faded to be replaced by the sameness of the mock invasions. It had, again, become war games. All of the dreams, all of the ideals, I had when I was so anxious to join the Navy seemed remote when involved in the repetitive practice. What I had forgotten is that repetition has always been the key to learning, to understanding, and most importantly, to doing when the situation presented itself.

That New Year of 1945 came and went barely noticed. We were

moored in Berth No. 2 at Sand Island, Honolulu. The only difference between 1 January 1945 and all the other days was that we had no maneuvers. It wasn't yet a time for celebrating, but still we had that one free day to reflect on why we were there. It was getting more and more difficult to remember that far out in the Pacific, the war was still going on.

A few days later, on 6 January 1945, we were doing tactical training exercises with the Transport Squadron 15. The USS *NAPA* zigzagged in and out and around the islands for three days, arriving back at Sand Island on 9 January. Sand Island was hardly more than just what its name implied—sand, dust and a few palm trees. It reminded me of the old cartoons of someone shipwrecked on a tiny island with one palm tree. It was hardly more than that.

For the next two weeks, we continued our practice with the Transport Squadron, and on 18 January 1945, we were back at Sand Island. I watched as the officers underwent gas mask instructions. I could still remember the stink, and vaguely wondered if the enemy gas just sought out officers and let us enlisted men alone. I recalled the gas chamber back at the Naval Training Station in San Diego. I wondered if the gas the officers masked up against was real or not. Just as I didn't test to see if it was the real thing back at Basic, neither did the officers there.

Near the end of January, the training was finally over and we were ready—more than ready—to get into action. By that time, I, for one, was getting tired of the repetitions. After all, how much more polished could we be? We knew our jobs and could now do them in our sleep, which was the idea. There would be no time in a real battle to ask questions or directions.

Garner, Estrada and I bunked near each other. One night, we were talking about the seemingly endless training. Garner said, "If I were any more polished, I'd look like an officer's boot."

Estrada and I laughed. We were both feeling the same way. "Think we'll spend the war here in Hawaii, going from island to island, firing at sea gulls?" Estrada asked.

"Sure seems like it," I agreed.

That was the way it was . . . after a while, it seemed overkill to keep practicing and practicing, without any foreseeable hope of putting that practice into reality. How eager I was to go to war!

CHAPTER 11
Enewitok

On 26 January 1945, a roll call was taken, and everyone was ordered to stay aboard. Maneuvers were suspended. All the landing craft was secured. I couldn't help but suspect that this was it. We'd gotten our orders! The USS *NAPA* was heading to war!

That morning, before 0900, on 27 January 1945, we finally got underway and left the Hawaiian Islands for the reality of war. On board the USS *NAPA* were the same elements of United States Marines whom we were ferrying to their ultimate destination, and ours. We zigzagged out of the island group. Destination: Unknown! At this point, I didn't care! We were heading out in the Pacific to engage the enemy—at last!

At sea, seven miles out, I was in the radio room when Col. Mustain of the US Marines, went on the public address system and announced: "Men, our destination is Iwo Jima, an enemy-held island in the volcano group. On D-day, 19 February, we will assault Blue Beach Sector. That is all." Everyone knew the rest! We were going to war!

Where the hell was Iwo Jima? That was the question asked and not answered. It had to be in the Pacific, but the island was as foreign and unknown to me as China or Outer Mongolia. It didn't matter! We were at sea, and we were, at last, going to war.

On board the USS *NAPA*, I was in the radio room, and the task at hand, one I'd practiced so many times it became automatic,

now was a serious routine. Messages were no longer mock communiqués; they were the telling of the war that was fast coming to
us. We had all learned our tasks well. All those practices, all those
repetitive exercises that had become routine were paying off. We
knew our stuff, and strange though it may seem now, I wasn't the
least bit nervous there in the radio shack taking down coded messages we called FOX and then decoding them, from other ships and
from the watchers on shore. All the coding and decoding I'd done
in Radio School and then later in the exercises in Hawaii had sunk
in and become a natural part of me. It was important, what we were
doing, but how important, we really didn't comprehend then. We
were off to find the war.

We continued under blackout as we headed away from the Hawaiian Islands in a convoy. There were no lights allowed that could
be detected by any enemy observer, no running lights on the ship,
and not even cigarette smoking on deck after dark. There was just
the silent, dark sea rolling under the ship, sending its spray on deck.
The dimmest of light could give our position away to the enemy,
and we were increasingly aware that the enemy could be lurking
anywhere, unseen in the vast expanse of the ocean. The other ships
in the convoy were shadowy images on the sea. We were now on
full alert, and every motion, every action, had to be deliberate and
carefully executed.

Even the painting of the ships was designed to conceal them.
The shades of gray blended with the waves of the ocean to cam-
ouflage them as much as possible from prying eyes of planes and
other ships or submarines. "Battleship gray" was more than a color;
it was a defense. We soon joined a convoy and continued on our
way across the ocean to rendezvous off Iwo Jima.

After a four-hour watch in the radio shack, even with the bunks
folded, I would sometimes dash down there and swing up into my
top bunk where I fit in the V fold. There I could take a quick nap
between shifts undetected. I was not the only one to discover this
way of snatching a fast nap.

Our lockers were just barely large enough to hold our sea bags.
There was one shelf for personal items that were kept in the ditty
bag. That was the extent of our individual belongings. Letters from
home, when we got them, were generally kept in the sea bag.

The showers were in one big room located on the same level as the bunks. No individual showers or privacy. At least ten at a time could shower under separate shower heads. No hot water either! There was no lingering over a shower.

The mess hall was the largest personal area of the ship, also located on the second deck. Coming into it, we would line up, get our sectioned metal trays, and go down the chow line until our trays were full. When there were no troops aboard, we had the pleasure of sitting down to eat. With troops, the tables were raised as there was no room to sit down for a meal. We all ate standing up.

There was a protocol in getting our meals. Those who were preparing to go on duty got to go ahead of the line. Those just coming off duty were regulated to the back of the line. Occasionally, a sailor or Marine who wasn't going on duty would try to jump the line. When this happened, the sailors, and Marines he cut in front of, would pick him up and pass him overhead to the back of the line. The interloper would be kicking and screaming all the way, but it did no good. Instead of being able to cut in, the guy would find himself deposited rudely at the end of the chow line.

What we didn't have aboard ship was any luxury. We made our own pasttimes, usually card games or writing home. When chow was over, some of us would lower a few tables and get up a game of poker. Chips were everything from matchsticks to poker chips brought from home or bought at some other port.

A lot of the troops on board, whether it be the ship's crew or the Marines, would go up top and lay on the deck with their shirts off to catch a breeze. Even then in early February, it was hot in the North Pacific, and the close quarters of the ship sure didn't help any in keeping cool.

We did not have what could be called, at the stretch of imagination, a "commissary." It was one small area, about five-feet by five-feet. There was just enough room in there for one man to sell cigarettes, candy, lifesavers, and gum. This was the ship's store.

Near the ship's store was where we left our clothes for the laundry room. They were left in bins, picked up and washed by the Stewards and brought back to the same spot, where we had to pick out our own clothes. Everything was marked with our last names, even down to our socks and underwear.

The officers had it comparatively easy, as they had their own quarters, two officers to a room, and were allowed to eat in there, too. They even had their meals served to them by our ship's Stewards. The officers did not mingle with us ordinary sailors.

In those years, the Stewards were all Black, and they were the food servers and personally saw to the officers' welfare. The officers did not eat the same fare as we did.

During our off-times, Garner and I used to speculate on what the officers ate. "Bet you steak and the trimmings," he ventured.

"Wouldn't doubt it," I agreed.

To this day, I have no idea what it was, but I knew it had to be better than what I ate in the ship's galley.

On 31 January 1945, we crossed the International Date Line, heading toward the Marianas, and then on to Iwo Jima. It was the only light moment in an otherwise grim reality. We erased an entire day, for on the other side of the International Date Line we emerged into, not 1 February 1945, but 2 February 1945. 1 February 1945 would be forever known to us as "the Lost Day." All of us aboard the NAPA became members of the Order of the Golden Dragon. The Order of the Golden Dragon reads: "All on Board were Duly Inducted into the Silent Mysteries of the Far East, having crossed the 180th Meridian on 31 January 1945 on Board the U.S.S. *NAPA* (APA 157) Signed by Golden Dragon, August Ruler of the 180th Meridian."

It may have seen incongruous in the midst of war to be celebrating the Imperial Domain of the Golden Dragon, but we took our moments where we could. I, for one, was ridiculously proud of my Golden Dragon card. After all, how many people can say they were catapulted into the future, if only by one day?

Three days later, on 5 February 1945, we anchored at Eniwetok Atoll in the Marshall Islands in the middle of the Pacific Ocean. It was the first anniversary of the taking of this Atoll by the Fourth Marines, the same unit we now carried into battle to Iwo Jima. The fighting a year ago had been fierce, but the Marines had prevailed and the Eniwetok and the Marshall Islands were now under the control of the United States Military.

Enewitok was a smoldering ruin when it was finally conquered. It would have been totally useless if not for the Seabees.

The Seabees were miracle-working steam-shovel operators, carpenters, plumbers, even mining engineers, and ditch diggers who could make an airstrip out of a muddy field or build a bridge that would hold tanks without so much as a load of cement, and all without blueprints. They had turned Eniwetok from an island with battle scarred palm trees and little else to a modern military base, with harbor facilities, roads, bridges, hospitals and air strips. They did it almost overnight with bulldozers, sweat, and their determined know-how. The Seabees, more properly known as the Naval Construction Battalion (shortened to CBs, which became more popularly known as Seabees), were in on every landing in the Pacific. They rebuilt what bombs had destroyed, and they built it out of scraps and imagination, and, they like to say, bailing wire and toothpicks. Whatever they built, we could be sure it was strong, stable and was done right.

Because of the Seabees, there was a decent and unlittered harbor at Eniwetok where the USS *NAPA* anchored to refuel. No liberty there. There wouldn't have been anything to see or anywhere to go, anyway. I, for one, was not too disappointed, even though I would have liked to have gotten off the ship for a few hours.

CHAPTER 12
Saipan

On 7 February, we left Eniwetok and headed for the last safe harbor before Iwo Jima: Saipan in the Marianas. The battle for Saipan was still fresh in the minds of many of the Marines aboard the USS *NAPA*.

Seven months earlier, on 7 July 1944, the U.S. Marines, 27th Division had taken Saipan, after a ferocious battle. The invasion of Saipan was a vital strategic step for the United States in order to defeat the Japanese. The Japanese, led by Commander Yoshitsugu Saito, were dug in and waiting for the Americans before launching their all-out attacks. In anticipation of an attack on the island, the Japanese placed bamboo sticks in strategic locations along the beach as sighting lines for their 105mm, 75mm, and 150mm guns that were on the high ground overlooking the beaches. The placement of the bamboo sticks was the same as though a flag man was on the beach signaling to the gunners where to fire. It was a simple, but extremely accurate, method of guiding the deadly fire.

Saipan was a new type of problem for the Americans. It was a large island of over 70 miles with terrain that changed from cane fields to swamps to cliffs and to the highest point, Mount Tapotchau. It took careful and precise planning and concentrated fighting by the Marines to win Saipan. When the Marines stormed the beach, the anticipated easy taking of Saipan turned deadly. Over 2,000 of the 20,000 Marines fell to the fire from the Japanese guns.

By nightfall, a large contingent of Japanese infantry, supported by tanks, attacked the left flank of the Marines.

The Japanese fought from caves, ravines and gullies, attacking the advancing forces from all sides. American artillery and tanks were rendered useless in the jungle environment of Saipan. The fighting became man-to-man with mortars, machine guns, and even bayonets.

The Japanese, desperate to save the island as a strategic port for their fleet and landing strips for their air force, had launched the infamous banzai, or suicide, attacks with terrifying success. It is difficult to defeat someone whose only aim is to die and take as many of his enemies with him as possible.

The night was suddenly filled with the shouting "BANZAI!" Screaming in unison, over a thousand Japanese charged all at once, running dozens abreast of each other. Japanese officers led the charge as they ran toward the Americans. Their swords drawn, swishing circles over their heads, firing their weapons, they screamed "BANZAI!" in one, continuous roar. Amazingly, among the rushing attackers were wounded Japanese soldiers committing their last act for their Emperor. As the American machine guns opened up on the charging hoard, sweeping back-and-forth across the sea of bodies, they kept coming, even to leaping or stumbling over the forms of their fallen comrades. The Americans continued their firing at the seeming impregnable wall of Japanese fighters. The barrels of the mortar tubes and machine gun barrels became so hot that they were soon rendered useless. Still, they came, storming toward the Americans, wielding their swords, their weapons clicking on empty chambers.

At last, the banzai attack was over, and around the Marines lay the bullet riddled bodies of the last of the Japanese defenders of Saipan.

As the end grew near, and it was clear that the island would be taken by the Americans, the Japanese Commander Yoshitsugu Saito committed hari kari by kneeling on a mat and falling upon his ceremonial sword. His top officers followed his example. It was a ritual no Westerner could understand, but was the honorable way for Japanese officers defeated in battle.

The tragedy of Saipan was fully recognized on 9 July 1944 when the Americans reached the northern end of the island. Thousands of the island's natives—men, women, and children—had leapt to their deaths over the cliffs rather than surrender to the Americans. They had been convinced by the Japanese that they would be brutally tortured if they were in the hands of the Americans. Rather than submit to the unspeakable atrocities they were positive would be visited upon them by the Americans, they chose suicide. There was nothing the Americans could do about it, in spite of having pleaded with them over a loud speaker that they would not be harmed.

In the caves of Saipan, the Japanese soldiers, who were also convinced that the Americans were barbarians who would subject them to unspeakable acts should they be captured, chose to take their own lives in a most horrible fashion. They pulled the pins of their own grenades, held them to their chest, and blew themselves up.

That type of behavior of both civilians and opposing military was not something that Americans could understand. It pointed out one vast and important difference between the two cultures and the two forces.

The terrible cost of Saipan proved worthwhile when the Americans set up a much needed supply depot, as well as an airbase, and a safe harbor for ships to refuel.

The victory of Saipan did not come easily or without heavy costs. Nearly 24,000 Japanese had died in the defense of the island, and almost 3,500 Americans had lost their lives there. It was a dear price to pay, and the reminders were still there as we anchored in Saipan Harbor on 11 February 1945. The United States Marine cemetery with its seemingly endless rows of white square grave markers bore silent testimony to those who had paid the ultimate sacrifice taking the island. Of all the reminders of war, of all the talk about battles to come, those mute white markers underscored the reality of what we would face. The history of the Battle of Saipan was a precursor to what we would soon face on Iwo Jima.

The Marines we were carrying onboard the USS *NAPA* disembarked at Saipan to pay homage to those who had fallen before. Some of the Marines who were in those graves had been buddies of the ones who now were going to Iwo Jima. It was a somber time. The Marines came back reinforced for battle. It had nothing to do with revenge, but with understanding the reasons why it was so important to continue to take the islands between Saipan and Japan, and the price it would cost to accomplish it. No one believed we would not be successful, in spite of the well-earned reputation of the Japanese as fighters who would not give up, even when they knew they could not win. They were a formidable foe, but then, so were the US Navy and Marine Corps.

There was a storm somewhere out in the Pacific, and its edge had reached Saipan, turning the ocean into high choppy waves. It was like riding a never ending roller coaster. The ship pitched and rolled as though it were a toy in a bathtub. The ocean was showing its might.

Forward and aft on the top deck of the USS *NAPA* were the small boats that would ultimately carry the Marine troops and the beach parties into battle. These were LCVP (landing craft vehicle-personnel), which carried approximately twenty-six men plus the coxswain plus the boat handler, and one LCM, which was larger than the LCVPs.

The following night, 12 February 1945, in spite of the unpredictable weather and the rough sea, the USS *NAPA* hauled anchor and headed out for one final practice run before the big game. Out beyond the islands, the Marines and the forty-six man beach party, including myself, climbed down rope ladders over the side of the USS *NAPA* and piled aboard the amphibious craft to practice landing in preparation for Iwo. Because there was no resistance, no enemy ships off our port bow, no enemy airplanes buzzing the ships, it wasn't quite the realistic dress rehearsal it should have been, but the commanders were satisfied with our night rendezvous and amphibious landings.

Unlike the practice landings on Maui, even with the battle scars still fresh in Pearl Harbor, these landings were different. We were now in the very shadow of war, but we were confident. We were going to Iwo Jima, and we would be victorious. I envisioned in my mind that the sands of Iwo Jima would be comparable to the sands at Maui.

The following afternoon, we were back at Saipan Harbor, congratulating each other on the success of our practice mission. For the next two days, we could try to forget a war was going on. Liberty at last, which meant we were able to get beer on shore. All preparations were underway to be certain we were stocked and ready to fight, this included taking on supplies such as food as well as ammunition. The ship itself was tested, its engines run and listened to by ears that could determine if they were off by identifying and categorizing their hum. We were at liberty, the last for many of the crew, but we couldn't dwell on that. If that was in the back of our heads, we tried not to show it, tried not to think of the white grave markers on this faraway island. Liberty meant a small respite from war, a breathing space before we set to sea again on our final lap to Iwo Jima.

What we could not know was that, regardless of how many times we practiced landings on sandy beaches, whether it be California, the Hawaiian Islands, or in the North Pacific in the Marianas, nothing could have prepared us for the black, shifting sands of Iwo Jima. Looking back, it was probably a good thing that we didn't know what lay ahead, didn't know that the beaches of Iwo Jima were unlike anything we had ever seen, or would likely to see again. It was good because, had we known, that last night on Saipan, none of us would have been quite so boisterous, quite so eager, or felt quite so prepared. A fighting force that isn't prepared may as well give up without a fight. That wasn't us, and that wasn't the Marines we carried aboard the USS *NAPA*.

CHAPTER 13
Iwo Jima

For two months before the assault on Iwo Jima, that small, 4.5-mile long island was unmercifully, and continuously bombed, until nothing alive could have existed on the island. Then for three days before we landed, the Navy ratcheted up the pounding of the island. Six United States Navy battleships launched a continuous barrage on the island, giving it all they had. The sound was deafening. The shelling of the island was accompanied by more bombing by American planes. The island was left barren; lifeless. There was no anticipation of any meaningful resistance after the fierce and determined effort to annihilate the defenders of the island.

In spite of all of the reconnaissance of the island, all of the bombing that had preceded the arrival of our convoy, what no one knew was that the Japanese were burrowed under the island in an endless honeycomb. They had long anticipated that Iwo Jima might well be their pivotal battle before opposing forces would invade their homeland, and they were prepared to defend it at all costs.

All of the surface bombing had no effect on their hidden burrows. They were determined to fight to the last man, the last bullet to defend what they fully understood to be the last bastion between their homeland and the approaching American forces. That tiny island in the Pacific was their last stand. It had two major air fields from which the Japanese could launch attacks throughout the Pacific, or protect their homeland. They definitely were in a position to send Japanese fighter planes to attack American bombers on their way to Japan.

In American hands, the island was ideally situated for the landing of crippled bombers returning from raids on Japan. There was no place safe for the B-29 Superfortresses to land, and they had no fighter escorts as fighters did not have range enough to fly to Japan and back. Many of the bombers that sustained damage to land on their way back from Japan had to ditch in shark-infested waters; few of the crewmembers were rescued from that fate. With Iwo Jima in American hands, the two airfields could be turned to life-savers for the bombers, and fighter planes could refuel there and escort the bombers.

Both sides had urgent reasons to triumph over the 2.5 mile wide island, approximately 650 miles south of the main Japanese island of Honshu. That lonely little, uninhabited island had become a significant and strategic cog in the wheels of battle for both sides.

There is something about war and danger that brings people closer to God. I doubt if there was a man on board the USS *NAPA* who did not pray, who did not realize the thin veil that separated man from his Maker. Religious services were held for Catholic Mass, Protestant Divine Services and Jewish Devine Services. There was not a one of the services that was not filled to capacity! Those who believed differently prayed in other fashions and held their own services, but they were just as fervent and sincere. There wasn't one among us who didn't realize that this journey to the small dark island in the Pacific might be our last, and tomorrow we might be meeting the Creator of All. Silently, without voicing it, we knew there was a good chance that our ship's company, and the compliment of Marines we carried, would be far less when we finished our assault on Iwo Jima. Victory that we all knew, and prayed, would be ours, could not come without a price. How heavy of a price, none of us could have anticipated.

That night, while the ship was sailing toward Iwo Jima on a sea that was still rough from the hurricane way out in the Pacific, each of us reflected on our own thoughts. I couldn't help thinking of home and family, of my brother Elton who was somewhere out there in a C46 on the Burma-China campaign, of my cousin Vernon Edison, whom I called "V.E." who had been shot down and killed over France. I didn't dwell on them, didn't think about the possibility that I might never see Elton again. I wondered that night how Elton was, and if he faced the same fears mixed with determination as I did.

Duty took over and all thoughts of home and family had to be shelved, stored back in a box in memory for another day. Right now there was a war going on, and the ships on either side of us, as silent and dark as ours, were steaming with us toward the objective—Iwo Jima. This was what all of that repetitive training was about. There was no time to think things out; training took over, and I did what was needed.

Until Iwo Jima could be taken, those two airstrips on it were the launching point for Japanese fighters who intercepted the B-29s on their way to Japan, and on to the Marianas themselves. The fighters were tenacious, and they had to be stopped. The Japanese fighters did not give up, easily, if at all. If they knew they were mortally wounded, they turned their blazing planes into fiery missiles and aimed them at the nearest American target . . . plane or ship.

For those three days before the projected landing, American warships pounded the island with their big guns. Bombers flew across the island, dropping their big payloads. Fighters came in low and strafed the island with their machine guns. There wasn't an inch of Iwo Jima that didn't feel the mighty force of the American fire power, from air and from sea. The two airstrips, while bombed, were carefully spared the worse of it in anticipation of their use by friendly forces after the island was taken. There was no doubt whatsoever that the island would be taken easily.

We anticipated that there wouldn't be much left of the island, or anyone left alive on it, by the time we reached its shores since for the previous seven months the sky over Iwo Jima had been black with bombs. This operation, this landing on Iwo Jima, was considered as a mop up. There was to be expected small pockets of resistance, but nothing big could have survived the pounding that volcanic rock had taken all those months, and the intensified shelling for the last three days.

Three Marine divisions, under the command of Lt. General Holland M. Smith, were scheduled to land. Our 4th Division Marines were eager to hit the beach and claim the island.

On 18 February 1945, thirteen of us from the beach party, including radiomen, signalmen, and boat handlers, were transferred

Arvy Geurin's Naval graduation picture.

Standing—Gregory C. Kiewetz, RM3C;
Seated: Samuel R. McMahan, RM3C, Thomas Fitzgerals,
CBM; Standing at back: Arvy A. Geurin, RM3C

Beach Party

FIRST ROW, *left to right:* Rouse, G. E., CBM; Lt. C. W. Reynolds, USN; Haedel, P. P. HA1c; Rouleau, J., BM2c; Buckles, K. W., HA1c; Streng, W. L., HA1c; Wright, N. E., SF1c; Neustedt, A. W., S1c; Pertl, C. S., S1c; McLeary, C. W., S1c; Carlson, G. A., EM3c; Gregory, H. F., F1c; Estrada, L., RM2c; Denney, D. D., PhM2c.

SECOND ROW, *left to right:* Otto, R., BM2c; Mueller, D. M., S1c; McCullar, G. L., MoMM2c; Hassell, J. W., F2c; Hoegen, G. A., S1c; Williams, M. T., S1c; Dearen, R. P., S1c; Potz, J. R., S1c; Moody, D. W., S2c; Ward, J. M., CM1c; Garner, W. H., RM3c; Brophy, J. F., SM3c; Lt. F. Skubitz.

THIRD ROW, *left to right:* Geurin, A. A., RM3c; Kapp, J. A., SM3c; Bowen, J. D., S1c; Bowen, H. L., MoMM2c; Utz, J. F., Jr., HA2c; Hornick, H. W., S2c; Livsie, E. A., Jr., HA1c; Eagen, M. H., SM3c; Spurr, E. C., HA1c; Staton, E. L., RM3c; Wallingford, J. W., CM2c; Jacobs, M. K., S1c; Mancillas, P., S1c; Wood, G. D., S2c; Corban, C. B., PhM2c.

Not in picture: Bennen, G. R., S1c; McNeely, M., RM3c; Reed, J. M., S1c; Lt. Comdr. E. duPont, Jr.

USS *NAPA*, APA 157

Top to bottom; left to right: Marines and Navy Beach Party preparing to go ashore. They are standing on the deck of the NAPA *getting ready to go over the side into a LCVP (Landing Craft, Vehicle Personnel); Supplies on deck getting ready to go ashore. Middle left: Wounded Marine being hoisted aboard the USS* NAPA *(February 1945); Middle right: Looking toward Iwo Jima from the deck of the USS* NAPA; *Bottom: LCVP in zigzag pattern going to Iwo Jima, loaded with Marines and Navy Beach Party, including Radiomen.*

Arvy far left; John "Jack" A. Kapp, Jr., SM2C, Weldon H. Woods, S1C

Showing damaged tanks and landing craft shortly after the beginning of the barrage from Mt. Surabachi, shown in the background.

At Guam, Marianas Islands. Top picture is seemingly endless line of ambulances waiting to take aboard the wounded; Middle left picture: Ambulatory wounded Marines and Sailors on the dock at Guam; Middle right picture: More seriously wounded being taken off the NAPA at Guam; Bottom: Marines wounded on Iwo Jima leaving the USS NAPA at Guam.

The USS NAPA *was rammed by the USS* LOGAN *during blackout off the waters at Iwo Jima. This picture shows the damage from the hull of the* LOGAN.

to an LST as recon to set up communications from shore to ship. The radios were bulky and required three men each to assemble and employ. There were four teams of us, under command of Lt. Frank Skubitz, who hailed from Ely, Minnesota, and Lt. Commander Ernest duPont from Wilmington, Delaware.

On board the LST, there was almost a lightheartedness that replaced the earlier realization that we were going into a dangerous battle. The scuttlebutt was that, since the island had been saturated with bombs and pounded by the big guns from the war ships, the operation would take no more than three days to complete and the airstrips would be ours! It was that type of talk that encouraged the kind of joking that, unless you are there, no one could understand. One of the fellows even speculated about capturing a Sumari sword. I laughed, because even I knew any Japanese fighters left on Iwo Jima would not be Sumari, and not likely to have that type of weapon with them, or left lying about on the island.

That night, aboard the LST just off Iwo Jima, the first order of business was to break down our carbines and clean and oil them. They were treated like our best buddy, and we made sure they were in proper working order. The carbine could make the difference between living and dying in battle.

When my carbine was ready, I spent the rest of the time writing letters to home. Since I knew our mail was censored, I had code words made up for Mom to let her know what was really going on with me. I was able to let Mom know where I was, without ever giving away anything that would have been censored. Long before I left home, I had studied the islands of the Pacific on the way to Japan and had "named" them with names of family and friends. For instance, Hawaii was called "Sharkey" after a good friend with whom I had gone to school. Each of the islands was named like that, with Iwo Jima, that tiny dot, called "V.E" after my cousin. So it was that night aboard the LST I wrote Mom, "I saw V.E." and she would know I was going to Iwo Jima.

Some time later, a picture from Iwo Jima appeared in *TIME Magazine*, that was taken by Joseph Rosenthal, war correspondent and photographer for Associated Press. It was a sweeping picture of the beach at Iwo Jima, and showed the "Marines" in their foxholes. Amazingly, Mom knew by looking at that picture that the"Marine" in the bomb crater looking toward Rosenthal's lens was me. Because of my letter with code word for Iwo, she knew I was there on that island at that time. All of us were dressed the same, sailors and Marines alike, but Mom knew it was me. Mothers always know, just as Harlon Block's mother knew it was he, and not Harry Hansen, who was one of the flag raisers of the second United States flag to go up on Mount Surabachi on 23 February 1945. Mom cut out that picture and saved it to show to me when I finally got home.

The talk was about souvenirs and bragging rights. It wasn't as inappropriate as it might sound these many years after the fact. At the time, it was what kept us from being scared. We'd go into Iwo Jima, fight a few skirmishes, pick up souvenirs from the beach, maybe jump into bomb craters and shoot our rifles, and we'd go back to the USS *NAPA* in triumph.

It seemed less and less likely that any of us would become unwilling permanent residents of Iwo Jima. We were assured that there would not be the heavy resistance that occupied our minds only days before. There could not be anyone or anything left alive on Iwo Jima that would pose a danger to the US Marines, or to us as the beach reconnaissance landing party. By then, I was actually looking forward to being on Iwo Jima, to experiencing my first taste of battle.

On 19 February 1945, at 9:00 a.m., our LST sat just off Iwo Jima and I got my first glimpse of the island. Mt. Surabachi, the volcano, rose high above the tiny swatch of the island, that had been formed eons ago by its eruption. It wasn't an impressive sight, the dome-like top of the volcano sticking out of the finger of an island. It surprised me to see it, so small and seemingly harmless, surrounded by a mist that could have been early morning fog but was more than likely the residue from the bombing and shelling.

This was it! This was the objective, the reason why we were here in the middle of the Pacific. The small island couldn't have been more than three miles in length, if that. How long would it take a determined force of US Marines to transverse the distance between shore and the top of the mountain? Not more than the three days rumored, if that.

The first two waves went in, and there was little resistance. There was enough for us to know that the Japanese were still there, but the comparatively mild fighting also assured us that there wouldn't a big battle here. It was a mop up, no more.

We went in on the third wave, and all hell broke loose.

What we did not know unit it was far too late was that the defenders of this island did not fight like their German counterparts! They fought relentlessly, burrowed under the island in an endless honeycomb, stayed at their posts to the point of starvation or the last bullet. They had dug in their defenses, and they had complete control of the highest point, Mt. Surabachi. Their commander allowed the first two waves to land without playing his hand. It was the third wave that felt the full rage and fury of the Japanese forces, and we began to understand that this was not to be a three day operation.

In the third wave, we hit the beach at Iwo Jima. All the ships, the battleships, the heavy cruisers, the destroyers, were firing on the beach. As we went in, I saw our fighters strafing the island, and bombing Mt. Suribachi. I couldn't see much because of the smoke and fire from the bombing and shelling. All I could hear was the terrible sound of the guns, from the 16-inch guns down to the five-inch guns. I could hear the burst of the 40mm guns. *BOOM! BOOM! BOOM!* A demonic rhythm filled my ears. The firing of the mortars split the air above me, but in spite of all of this, it didn't dawn on me that I could actually be killed. I saw a couple of bursts out in the water near our landing craft. Our amtrack hit the beach, and we couldn't go any farther in the volcanic ash. I jumped over the sides with the Marines and the beach party, and ran up to the first terrace and lay down, waiting orders. I'd expected sand, not this slippery black ash that wouldn't even allow the digging of protective fox holes. It was impossible to gain a foothold. Footprints were quickly filled with the shifting, gritty, fine ash.

The beach had been bombed for thirty days by heavy bombers. It was made of black, volcanic ash. Craters made by incoming shells or lobbed from Mt. Suribachi by the Japanese were the only cover. There were no trees, no boulders, no buildings, no natural cover. There were bodies of dead Marines lying all around. The air was black with the flack from the ships and the return fire from the Japanese who were dug in above us. It was nothing like I expected! It was a living nightmare! The firestorm was as thick as raindrops during a monsoon, and as deadly as an acid bath.

When I hit the beach, I realized that all of the practice invasions we made the year as we prepared for this actual landing were to no avail, not only because of the onslaught of continuous mortar and machine gun fire, but also the volcanic ash, which was nothing like the sands at San Diego or Maui or the islands of the Marianas. The machine gun fire, the ship's big guns blazing, and all of the other chaos on the beach did in no way resemble the orderly, choreographed training we had undergone, back in California and again on Maui. There had been no way to anticipate that actuality of what we faced that day on the beach of Iwo Jima.

With all the firefights there, all of the heavy guns firing over our heads and at us, from behind us and above us, it was a miracle that I didn't die there. A war correspondent said it best when he said, "If you were on the beach at Iwo Jima and didn't get hit, it was like being out in a rainstorm and not getting wet!"

Iwo Jima. Shaped like a misshapen pork chop, that small island in the Pacific that no one outside of Japan had even heard of before that day in February 1945, would become known as "*The bloodiest Pacific battle of World War II where uncommon valor was a common virtue.*" (Attributed to Admiral Chester W. Nimitz, US Navy.) When it was over, the toll would be gruesome: 6,821 Marines killed; 19,938 wounded, and 20,000 of the 21,000 Japanese burrowed in there were dead.

Yet, in spite of the awesome price, the cost of the island saved countless lives. When Iwo Jima was secured, over 2,500 B-29s made emergency landings on the island's two captured airfields, a feat not possible before the taking of the island. Until then, the 11-man crews of the wounded B-29s had only one option: ditch in the Pacific and take the chance of being rescued before being eaten by sharks or killed by the enemy.

At the time of our landing, we weren't thinking of B-29s or their crews. Our main task was to take the island, secure the airfields, and stay alive long enough to do it! Later, we would realize how important that battle was and why. The first B-29 to crash-land at Iwo Jima after the airfields were secured made that statement better than any field commander or war strategist could!

Much later, after the battle was well underway, we realized the cunning of the Japanese commander at Iwo Jima. He had allowed the first two waves of Marines to land without much resistance, giving false security to the ones who followed. On the third wave, the Japanese cut loose with everything they had. It was clear that their intent was to kill every living creature on the beach—Marine and Navy alike! All of us on the beach were like ducks in a rain barrel waiting to be slaughtered!

At the time of our landing, the 8th Marines Ammunition Company also landed, bringing with them much-needed ammunition of all calibers and types for the fighting forces. These were Marines from Montford Point, near Camp Lejeune, North Carolina. They were all Black, and they had the hazardous job of getting ammunition from ship to shore. They did their job under fire, and knew full well their ammo dumps would be prime targets for the enemy. Yet, they did not shirk their duty, and as a result, several of these Marines were wounded or died there on Iwo Jima. The rain of hell that fell down from Mount Surabachi did not discriminate. Before war's end, over 900 of these brave Marines from Montford Point had given their lives.

The heavy saturated bombing of the island in the weeks before the invasion did not deter the Japanese. They had dug in under the island's volcanic crust. They had constructed 16 miles of interlocking tunnels 46-feet under the island in a virtual rabbit warren or labyrinth. They served not only as defensive positions, but also as a hospital and living quarters for the Japanese defenders. When attacked, they simply ran or crawled from one tunnel complex to another to avoid the bombs.

The beach at Iwo Jima was nothing like I pictured. There was no sand. There was just black, slippery ash that had been spewed for hundreds of years before from its highest point, Mt. Surabachi, which was still an active volcano. At that time in February 1945, it was recovering from a phreatic eruption just three months before in December 1944. While eruptions were not fierce, they were enough to know that the volcano did not sleep.

The volcano was shrouded in a smoky haze from the firing of shells and fog that day. Over 525 feet above the island, top of the mountain stuck out like a ghost above the haze. Two waves of Marines, carried in by the amphibious boats, had landed before we hit the beach. Although the Japanese had yet to let loose with the full force of their fire power, they had fired down from Mt. Surabachi onto the beach below. The beach was already cluttered with the bodies of those Marines who did not make it up the first of the three terraces. Amphibious boats with shell holes the size of a Volkswagen lay on their sides, crumbled and useless.

When I tumbled out of the amphibious boat with the Marines and the rest of the beach party, it was all together in a surge. There were forty-six of us in the boat, and the jostling of the waves and shifting of human bodies against each other made it impossible to keep track of each other. I scrambled onto the beach, holding tightly to my section of the radio.

I was part of a three-man crew who operated the radio. The radio came in three parts, each of us having the responsibility for a section. It took all of us to assemble it. Once we were on the beach, the three of us were separated. I had to find the other two in order to be able to put our radio together and run up the antennae. It was what we all trained for; it was the reason I was on this beach amid the fire and hellstorm.

I jumped into the nearest bomb hole, which served as a ready-made foxhole. Garner and Estrada, the other members of my team, were nowhere near me. Somehow in the mad dash for the impromptu foxholes, we had become separated. I landed in a bomb crater with Frenchie LaRue, Boatswain Mate 2nd Class. I heard the distinctive whine of high power shell and instinctively put my hands over my helmet and ducked. The whizzing shell slammed into the shallow foxhole next to me. After the explosion, I glanced where it had landed. Two of the men from another APA with whom I went to Radio School—Nowak and Novac—had run up their antennas right near us. The Japanese zeroed in on their radio and killed both of them. I didn't see the third member of their crew. I could only pray that he had made it out of that bomb hole alive. That was the chance all radiomen knew they took. It was a grim reminder of how both important and dangerous our mission was. Nevertheless, we had the critical job of shore-to-ship communications. The very real possibility of death on the battlefield was one of the most important reasons why there were multiple radio crews hitting the beach with the Marines.

The war correspondent and photographer, Joseph Rosenthal, was running up and down the beach, snapping pictures. I thought he was crazy. He was unarmed, and seemingly unaware of the danger. He was bending over, with a camera held to his chest. He kept snapping pictures as he ran. He took a sweeping picture of the beach and happened to catch me in a bomb crater looking up at him. The picture later appeared in *LIFE Magazine*. Rosenthal paid no attention to the shelling, concentrating instead on recording the action with his camera. It was Rosenthal who would later take the now famous photograph of the raising of the flag on Mt. Suribachi, that would become an icon and symbol of victory over Japan. That was four days away, on 23 February 1945.

God is alive in a foxhole! I turned to Frenchie LaRue, and told him that I believed in God. "If something were to happen to me," I said urgently to LaRue, "notify my parents. Talk to them. Tell them what happened."

"Sure, Geurin," he agreed. "Do the same for me."

Less than ten minutes later, hot shells rained down on us. They whizzed by over our heads. Hell had broken loose on Iwo Jima. I ducked as low as I could in the shallow bomb crater, and lowered my head, praying the shells would go over us. I heard and felt the dust and debris churned up by the live fire as it hit my helmet. It felt and sounded like machine gun fire pinging off my helmet. When the firing had ceased for the space of a breath, I looked over to LaRue, and his arm was hanging by a thread of skin. Blood poured down his arm, making a muddy river in the dust and dirt that covered him as it went. He was apparently not even aware he'd been hit.

I yelled at him, "LaRue! You've been hit!"

Stunned, he grabbed his arm with his good hand, the blood instantly seeping between his fingers. He handed me his walkie-talkie, and scrambled out of the bomb crater. He ran down to the beach behind us. I learned later that he had caught a landing craft that had the ramp almost all the way up, in preparation of heading back to the ship. He dove, hit it with his stomach, and toppled over into the bottom of the boat. That was the last I saw of him. I heard that he had been taken out to the hospital ship, and his arm was saved.

The walkie-talkie squawked. The Lieutenant called Frenchie LaRue by his code name. "Geurin here, sir, LaRue was wounded. Returned to the ship." I answered.

"Geurin, you set up yet?" The Lieutenant asked.

"No, sir!" I answered, breathing very close to the walkie-talkie, not wanting to be overheard. In retrospect, that was more instinctive than productive for the roar on the beach that day was so high even a shout through a megaphone couldn't have been heard.

"Separated from my crew."

"Get that radio set up, Geurin!"The lieutenant told me."Find the rest, now!"

"Yes sir!" I slung the walkie-talkie over my shoulder, grabbed my section of the radio and crawled out of the bomb crater.The beach was cluttered with large, spent shells, some still hot, discarded rifles, and the twisted and bloodied bodies of the fallen Marines.The shell holes and slippery volcanic ash made for rough going. It was impossible to run in the black sands of Iwo Jima. I stumbled over a body of a Marine half submerged in the sandy ash, who had been all but decapitated, his head at an impossible angle, blood coming out his nose and caking his uniform. His rifle lay slung out beside him.A grenade was clutched in his fist. I moved quickly away from him and tried to run, hunched over, zigzagging where I could.The terrace was sloped, and the black ash was like trying to get traction on marbles underfoot. For every step forward, it seemed that I slid two steps backwards!

Mortars were flying everywhere! By the time I could hear the whine of the mortar, it was too late. However, instinctively, I dived for cover whenever the sound buzzed by. By this time, I was next to a Marine whom I knew was an officer, although he didn't have any distinguishing bars or other symbols of command. It wasn't unusual for officers on the battlefield to remove, or conceal, the symbols of their ranks, for those bars and stars were immediate targets for the enemy. I still had not found the rest of my radio crew. "Sir," I said urgently,"Radioman 3rd Class Geurin. I can't find the rest of my radio crew. Should I continue looking?"

He told me,"Son, you have an order. Carry it out!"

I ran up and down the beach, trying to find Garner and Estrada. Torn and bloodied bodies of Marines dotted the terraces. They were valiant men of the Fourth Marines who paid the ultimate price. During my dash down the beach, I heard the telltale whine of a mortar and dove for cover into a fortified pillbox where the Japanese had once had supplies.There had to be at least ten Marines in there. When I first went in, it was dark, and I didn't know if there were Japanese or friendlies there, so I started to back out, rifle at the ready. I wanted to get out, and get out fast! Someone grabbed me by the nape of my neck and yanked me back.All I can remember is his voice saying in English,"You leave after we do!" In other words, he didn't want me to give away their position. I stayed put! Thank God they were Marines in there, not Japanese!

When there was a break in the shelling, the Marines in the shadowy cave of the pillbox took up their rifles and emerged, firing onto the terrace.

After searching for what seemed an eternity, but was probably no more than fifteen minutes, the walkie-talkie squawked. It was the Lieutenant, "Geurin, got the radio assembled?"

"No, sir!" I replied.

"Forget the radio," he ordered. "Grab whoever you can. Get a stretcher and go up to the airfield. We got wounded up there. Need to take them back to the beach for transport. Do it now, Geurin!" He told me where to go to find the stretchers the medics had brought on shore.

"Yes, sir," I answered. Thus it was that my radio was never assembled on the beach of Iwo Jima. Along with the others, at the pre-designated spot, I grabbed an empty, folded canvas stretcher and Seaman 1st Class Weldon Woods, a 17-year old, and I made our way up to the airfield. We zigzagged our way up the terraces, ducking each time mortar fire zinged by us.

The ships at sea were still shelling over our heads, and deadly fire came down from above us like red hot rain. It was a hellstorm of fire from all directions. The Japanese lobbed shell after shell down from Mount Suribachi, and the Marines and Navy fired back just as rapidly. All around us the black ash was flung upwards under the relentless combination of boots scrambling over it and live fire hitting it. The result was a black rain pouring over us.

At the airfield, Woods and I grabbed a wounded Marine and tumbled him onto the stretcher. We didn't waste any time being gentle, and turned around and started running back toward the beach with him. I was on the front end of the stretcher. The involuntary moan from the Marine as he was jostled with our running told me that he was still alive. The urgency to get him down to the beach and comparative safety was underscored by the live fire hitting near me.

Woods yelled from behind me, "Hurry! Hurry! They're shooting at us!" I looked around and, I swear, he was running on his knees, carrying the back end of the stretcher until he could get up on his feet. Machine gun fire hit the ground right behind him. Puffs of dust, dirt and ash engulfed his feet.

We somehow made that zigzagged trip down to the beach and not-too-gently lay the stretcher on the muddy and cluttered shoreline, in a row with numerous other stretchers filled with the wounded who were waiting their turn back to the ship. They were exposed, those wounded Marines, to the constant fire from Mount Suribachi, and snipers who briefly darted out from their holes or sighted down from a concrete pillbox. The Japanese zeroed in on the wounded, and some of the now-helpless causalities, lying in stretchers on the cluttered beach became part of the fatality statistics that began to pile up on Iwo Jima. The landing crafts that came for them were also targets for the fierce fire power from Mount Suribachi, or the snipers and machine gunners in the pillboxes.

The Marines did their best to neutralize the fire power from the pillboxes by using flame throwers and tossing hand grenades in the entrances. I could hear the screams of the Japanese who were caught in the fire from the flamethrowers. Their writhing screams of agony pierced through the whooshing sound of the flame throwers. Some of them stumbled out of the entrances to the pillboxes, still on fire, their screams meaningless words of unimaginable pain before they crumbled into a charred and unrecognizable heap. Alongside the flame throwers, other Marines took a firm stance and fired repeatedly into the pillboxes to ensure that the Japanese in there did not have the opportunity to disable the flame thrower before his job was finished. It was horrible to hear and watch, but it was the only way to neutralize the pillboxes and clear the way for further advance. Although this did eliminate some of the Japanese fighters, it had minimal effect at first, since the Japanese simply moved back into their tunnels to emerge at another hole or pillbox far away from the flames.

Later, our tanks, equipped with flame throwers, plowed through the terraces and blasted the pillboxes, and rabbit holes (those crevices the Japanese used to dart underground into the rabbit warren-like maze under the surface of Iwo Jima).

Woods and I didn't stop for a breath. There were more of the wounded Marines up there near the airfield who needed to be brought down to the beach, and at least be given the chance of survival. The firestorm never eased; in fact, it increased as the day wore on. There was no time to worry about becoming part of that ever-growing number of the wounded and dead on Iwo Jima. My mission had changed from shore-to-ship communications to helping save the lives of the brave Marines who had fallen in battle. I had never trained to be a stretcher bearer or a medic, but there on that piece of the volcanic island in the middle of the bloodiest battle in the Pacific, that is what I did and what I did with a growing sense of purpose. My arms strained with the weight of carrying a laden stretcher, my muscles protested, but there was no time to stretch or relieve the pressure. Pulled muscles in my arms were nothing compared to the gaping wounds I witnessed in the Marines we took down to the beach.

Woods hung right in there. His breathing, like mine, was labored from darting up the slippery slopes to the airfield, and rolling an injured Marine unto the stretcher, then running in a zigzag down to the beach. We did not pick and choose our "passengers." Some were very heavy; all were in full gear. They were bloodied and torn, some with arms dangling or feet blown away. Some with gaping throat wounds, or with their faces half-gone. Some could stumble to the stretcher, semi-conscious. Others were barely breathing as we maneuvered them onto the stretcher. The corpsmen had given emergency aid to many, but there were still those who held their own wounds, with blood seeping between their fingers. There were those whose blood bubbled through their wounds and I knew they would not make it, but we had to try. Those who were ambulatory had to make it down to the beach on their own, and they did, wounded helping wounded. The Marines were putting up a mighty battle for the airfields, and paying a high price for it.

On one of the seemingly endless trips to the beach, carrying a wounded Marine, I saw my friend, Harold Hornick, Seaman 2nd Class, lying dead on the beach. If I hadn't known for sure it was him, I wouldn't have recognized him—not because of his wounds, but because of the awful color he had turned. I was told it was Hornick, and I stopped for just a second to stare at him, then go on. That wasn't the time for personal mourning of the loss of a friend and shipmate. But it did bring it home, that this war was indiscriminate in who it took. I learned later that Hornick was buried on the same island where he had lost his life in duty to his country.

By this time, the beach was crowded with equipment, men, and the dead and dying. One could hardly move without stepping on a wounded Marine or sailor. There were amtracks turned on their sides from where they had been hit by enemy fire. Jeeps and tanks were abandoned on the beach, unable to get traction in the slippery black ash. The waves washed up on shore helmets, rifles, and fully loaded backpacks, all mute evidence of those who did not make it to land. Occasionally, a body would float to shore, carried by the harsh waves.

We set the Marines off the stretchers onto the beach to wait for transport back to the ship. We took off back to the fighting to gather more of the wounded. The wounded weren't safe on the beach because of the firepower being rained down on them. It was all we could do for them, and pray the transports would be on time. There was no time to stop and think about what we were doing. We just had to do it. This lasted all day to the night. Next to one of the wounded Marines was a face-down body of a Marine who had not made alive to shore, but washed up there by the powerful waves. There was not time to move the body away from the wounded. That wounded Marine had to lay there next to a dead comrade until transport came and took him back to a ship.

The night was illuminated with the light from the erratic bombing. No Marine or sailor moved on the beach at night, no more than there had to be.

An eerily glow of fires, the red hot explosions from both the ships and the Japanese up on Mount Suribachi fractured the darkness. An ammunition dump exploded on the beach, setting off a fireworks display that rivaled any stateside Independence Day celebration. Machine gun fire peppered the night air, the white hot flash of their muzzles a bright light in the night. Nighttime on Iwo Jima wasn't for sleeping, even for the most weary or battle-worn.

In the night, the Japanese crawled out of their rabbit holes and charred pillboxes and quietly overran Marines and sailors alike who were keeping watch in their foxholes or bomb craters. With a swift and quiet slash of their knives or bayonets, they dispatched our men with barely a sound. They were impossible to see as they slithered over familiar territory; only startled cries in the night told us that another Marine had fallen prey to the stealth of the Japanese. Now and then, we heard the cry out of a Japanese soldier who wasn't successful in his covert mission.

That night in a bombed out crater of black ash on Iwo Jima was the longest of my life. There was no sleeping for me. I heard the cries of Marines pierced by bayonets and the call of "Corpsman! Corpsman!" Until the voices weakened and there was a brief spat of silence. I could do nothing but sit scrunched in my foxhole. A Marine was in the foxhole with me. He put his finger to his lips and whispered, "Do not move. You don't want them to know where to fire, do you?" I shook my head. "Dead, you are useful to no one, man, remember that." I knew he was right. Hearing the voices crying out in the night, and not being able to do anything, was the hardest thing I have ever done. I had to sit there in that small, smelly hole and wait for dawn to edge over the horizon, and try to ignore the cries, and blank my mind from what they meant. The Marines on that beach were well-trained and in spite of temptation to do otherwise, they stayed put; I could only hope they could see in the dark.

Some corpsmen who answered those cries in the night were later found with their throats sliced. It was not uncommon for the enemy to call out, mimicking a wounded Marine, to lure a would-be rescuer into a deadly trap. At night, even with the sky lit by the flare of bombs and mortars, it was hard to tell if the person calling for assistance was a Marine, or if it were a trap. The corpsmen were duty bound to help whenever there was a cry for help. Many died as a result of their unflinching duty. Thirty-eight percent of the corpsmen on Iwo Jima fell victim themselves to enemy fire while trying to save the wounded. By the end of World War II over 1,250 corpsmen had given their own lives in attempting to carry out their lifesaving duties.

Hospital ships weren't immune, in spite of very large red crosses painted on them in very visible places that could be seen from air and sea alike. They, too, while caring for the wounded, were hit by enemy fire, including kamikaze attacks. These ships had no defense, and sustained severe damage and casualties of both staff and patients. The USS *COMFORT*, an American hospital ship, was to suffer that fate off Okinawa, just a month after the battle for Iwo Jima. Although the ship survived, nurses, doctors and wounded were killed in that attack.

That long night, I thought of my mother back in Bakersfield and how she would react if I didn't make it off this island alive. Would news of my death be met with the same kind of mind-bending horror that Aunt Lulu experienced when she learned her son V.E. would never again come home? I could only imagine how Mom felt when she learned Elton had been shot down. Try as I might, I couldn't get those images from my mind. The explosions that lit the night, the cries for help, coalesced in my mind. I deliberately thought of other things, of the bakery I would one day open, even without V.E. Somehow, without V.E. it just didn't hold the same attraction as before. I wondered if Uncle Vernon had lost his zeal, too. Death changes everything and everyone.

As dawn lit the flak-filled sky, I mentally readied myself for another day of stretcher bearing, of darting up the hilly terrain, fully exposed, to retrieve more wounded Marines; or to go from foxhole to foxhole to tend to those who had fallen prey to the night time raids. It may not have been what I came here for, but it was, nevertheless, an important and crucial mission to get our men off the island and to safety. The hellfire, that had never fully stopped during the night, began again in full force with the coming of daylight.

Out of the 46-men beach party from the USS *NAPA*, there were 13 wounded and two who died there at Iwo Jima, Seaman 2nd Class Harold W. Hornick being one of them, and Seaman 1st Class John Max Reed the other. Later, both of them would be buried at sea with honors. Others from the NAPA also lost their lives there: Lt (jg) Ernest J. Ritchie of Winchester, Massachusetts; Fireman 1st Class Anthony Morrone, Seaman 1st Class James Carlton Owens, Seaman 1st Class Benjamin Schlabach, and Lt (jg) Ford Eshleman, originally from Lake Mills, Wisconsin. They selfishlessly and without hesitation performed their duties to the United States of America and the US Navy. They fought with pride, and died with honor.

On the morning of 20 February at 0900, the remainder of our beach party rendezvoused at a predetermined point. I made it there, stretcher in hand, fully prepared to go again into the fray. The fighting was already more intense than yesterday. Waiting for us was another beach party from a different ship. We were free to go, to return to the USS *NAPA*. I was weary, bleary-eyed, and mind-tired. The images of what I had seen would never go away. I climbed into the amtrack with the others, and watched as the view of the island receded. The sound of the fighting seemed louder the closer we got to the ship. There was an entire convoy out there, lobbing shells onto Iwo Jima. Near the shore, the coxswain expertly weaved his way around the clutter. I saw bodies of Marines floating face down, and understood we could not stop to retrieve them because it was too late for them. The danger to our craft, and the personnel it carried, outweighed the humanitarian task of collecting the dead. There would be others who would see that each of the fallen was located, identified, and returned to their ships.

The rope ladder was over the side of the USS *NAPA* when we finally got there. The boat handler steadied the LCVP as I scrambled over the side and grabbed the ropes to climb aboard. It seemed a very long way up to the deck. Hands reached over the side to help us the final feet to the ship. I was never so happy to see the USS *NAPA*. For a wild moment, I felt like flinging myself down and kissing the deck! Thankfully, that moment quickly passed. I could still taste the sulfur that filled the air back there on the island. It was a taste I wouldn't soon forget.

When I reported to the radio room, jaws dropped in surprise. "Geurin," one of them said, "you're supposed to be dead!" It was then that I learned there had been a report that I had died on the beach.

"I'm glad you're wrong!" I shot back.

My fellow radiomen gathered around and showed me the reports. Other erroneous messages had been received, also, regarding the dead and injured. I also learned that no one had been in charge there on the beach. Our officers had been wounded and returned to the ship. Lt. Commander Ernest J. DuPont was wounded that first day, as was Lt. Frank Skubitz . The Chief Petty Officer also had returned to the ship before us, and the next in command was Frenchie LaRue, who had been severely wounded that first morning and returned to the ship. All who were left ashore were the radio operators, signalmen, and boat handlers—no officers. We returned that next morning, unaware that we had been without leadership on the beach. The confusion on Iwo Jima's beach was such that it would have been impossible to know what was happening anyway. There was no one in command to coordinate the group of radiomen and signal men, but it says much for our training that we did our jobs anyway, and walked through the fire of Iwo Jima to come, at last, back aboard the USS *NAPA*.

The USS *NAPA* moved out of the waters off Iwo Jima to what it called it's night retirement area, out of the reach of the guns from Iwo Jima, and supposedly safe from the kamikaze attacks. The kamikaze were those one-way suicide planes that were aimed at ships at sea. In the Japanese warrior culture, it was an honor to die that way, taking as many of their enemies as they could with them. We had what was left of the ship's complement and the injured Marines aboard.

CHAPTER 14
Collision

On the morning of 21 February 1945, at 0444, a ship was observed bearing down on the port bow of the USS *NAPA*. There were no running lights on either ship. Our navigator attempted to maneuver the ship to avoid the other vessel, but was unable to do so. All the crew could do was to watch helplessly as the dark ship came closer and closer. The slight turning of the USS *NAPA* that was managed turned out to be a literal lifesaver. The vessel hit portside with a resounding crash, and metal groaned as it buckled and tore under the force of the collision.

At 0446, I was bounced out of my bunk. The emergency blasts echoed over the loudspeakers. "NOW HEAR THIS! NOW HEAR THIS! MAN EMERGENCY STATIONS! PREPARE TO ABANDON SHIP! NOW HEAR THIS! NOW HEAR THIS! PREPARE TO ABANDON SHIP!" The warning was repeated. Before the last word died, I was in my uniform, including my flak jacket, running down the aisle to the hatch, tugging on my helmet as I ran. My only thought was that, somehow, a Japanese ship had snuck through our lines and we were rammed, or that a night flying kamikaze plane had hit its mark.

Up on the deck, there was fury of activity. The dark hull of another ship loomed ominously up against ours. Fire hoses had been dragged out, even though there was no fire, as a precaution. Hurried voices overlapped each other, but I heard enough to know that no enemy ship had slipped through the convoy to target the USS *NAPA*. The ship that rammed us was identified as the USS *LOGAN* APA 196, one of ours from the convoy. It hit the port side, just aft of the superstructure. In the pitch blackness, the other ship did not see the USS *NAPA* until it was too late. That was little comfort, for I soon realized we could still be sunk, and by one of our own. What irony!

The collision siren kept blasting! Then the authoritative voice shouted over the loud speaker once again, "NOW HEAR THIS! NOW HEAR THIS! ALL HANDS, NOW HEAR THIS! PREPARE ALL CASUALTIES TO ABANDON SHIP. PREPARE ALL CASUALTIES TO ABANDON SHIP!"

My blood ran cold. We had around 300 wounded Marines aboard, many of them in casts, some missing a leg or an arm, none of them capable of swimming in the dangerous waters off Iwo Jima to await rescue.

A Lieutenant—in the dark, it was impossible to identify him—ordered, "Down below. All of you. Get the casualties. Bring them topside. Hurry! Hurry!" The ship was badly listing to port. The USS *LOGAN* was still embedded into our hull.

I ran down into the ship to help get the casualties and take them on desk in preparation of abandoning ship. Some of these same Marines were the ones whom Woods and I had carried on a stretcher from the airfield on Iwo Jima. After all of that, after what they had suffered and Woods and I had done to help, they might die here, a victim, not of a Japanese bomb or machine gun, but of a collision with a friendly ship. I tried not to think of that as I helped a wounded Marine to his feet and together we made our way topside.

Evacuation plans had been made and drilled into us long ago. When "ABANDON SHIP!" sounded, it was the signal to jump overboard, and take with us whoever could not make it on his own. At the time, those were presumed to be wounded crewmembers of the USS *NAPA*. But now with war casualties, the plan remained the same, only all the more dire.

The plan shifted slightly, and now each of us to jump with a wounded Marine, with the goal of keeping the wounded in sight and afloat. The waters around Iwo Jima were shark infested. I knew there would be little chance of survival for either the Marines or any of us if we were in the water very long, for blood drew sharks and most of our casualties either had open wounds or bloody bandages. It would be like chumming for sharks! The other problem was that those in casts would surely sink to the bottom long before one of us could leap overboard to help. Our only hope if we went overboard was for one of our own vessels to rescue us, and as dark as it was at that hour of the morning before dawn, it was very likely most would be lost before a rescue effort could be launched.

All we could do, if the word came to abandon ship would be to toss them overboard, and pray they would survive and eventually be rescued. It was not a pleasant thought, but a necessary one. The corpsmen dispensed bottles of whiskey to the wounded Marines in preparation of their being dumped over the side. They knew their options, and prepared themselves. All of our landing craft were still in the water, back at Iwo Jima. There was no other way to save even a portion of the casualties and ourselves, other than jumping overboard.

With the wounded on deck, the USS *NAPA* lurched again, then shuddered. The prognosis wasn't good. I stationed myself alongside a Marine in a cast and waited for the order to go overboard with him. With luck, I could call on my swimming skills and keep both of us afloat.

Years later, I would learn a very personal story of that early morning from one of the Marines who had been there. He knew that at any moment, he would be in the water, fending for himself. His name is John William Marsh, and he was with the Fourth Marine Division. Many years after that fateful morning, he wrote the following letter when he read about the USS *NAPA* during a reunion of Iwo Jima survivors:

"We were aboard a LST! On the morning of 19 February, we proceeded to tank deck and got into designated tank and disembarked to form the 5th wave! There was no 4th wave, as we were told. It was a dummy wave. As the Japanese were in the custom of opening up on the wave to cut off the 1st to 3rd waves! We landed next to a Japanese black ship! We were between 2nd level and the 1st airport in a shell hole. Myself, Walter Liss and Donald Sheltton, when we got hit the 1st time by a small artillery or mortar shell. Liss seemed to be wounded the most. I was reaching for my gun belt to get sulphur to put on his wounds when it seemed like seconds later another shell came in, killing Liss and Shelton and

wounding me severely.

"*Corpsmen were very busy! A wounded Marine passed by and saw me and went to the aid station by the Mt. S, got another Marine, came back with a stretcher to get me! They proceeded to aid station. I have no idea how long we were there when we spotted a wave nearing the beach.*

"*They picked me up and ran down the beach where water was breaking on shore! I have no idea who they were, but I know I owe my life to them! As we boarded, I think was a Higgins boat, they told me the aid station was being shelled!*

"*We proceeded out to what I thought was a LST, I don't know! They took me aboard, dressed my wound, then transported me to the APA NAPA! Once aboard, they rushed me to the operating room as one of the severely wounded! As they rolled one patient off the table, they rolled me on.*

"*When I woke up, I was in a cast from my neck all down my body to toes of both feet. My foot was amputated! I was told by a corpsman. The doctors gave up their bunks to the wounded and were operating around the clock!*

"*I have no idea how long I was on the NAPA when all of a sudden, I heard a large blast and the ship was rocking! A Chief was running through the quarters. I yelled to him that I couldn't move! He got a stretcher, picked me up, put me on it, got another sailor and carried me up to the upper deck, starboard side of the bow. I waited to find out if the ship was sinking!*

"*I was informed they ran out of "May Wests." (Life jackets) It was finally decided that the ship was going to stay afloat. It wasn't until the next day that I learned they were close to throwing me overboard with a crewman jumping alongside of me with a flashlight until we got picked up. I know deep in my heart, he would never have found me. I would have sunk to the bottom with that cast on my leg! I owe a lot to so many hero sailors on the APA NAPA!*

"*We were finally put in an escort heading to Saipan. When we got there, we were told we were heading for Guam. Saipan was filled with wounded. I finally ended up in a base hospital in Guam. From there, I was sent to Pearl.*

"*I thank God that the NAPA never went down.*
Marine Corp John Marsh."

The USS *LOGAN* backed away, metal groaned against metal, then the USS *LOGAN* came forward again when the USS *NAPA* listed even more. It was apparent its withdrawal would have an even more disastrous result. The USS *NAPA* was listing 10-degrees, and the USS *LOGAN* stayed alongside to render assistance.

Repair parties worked feverishly shoring up watertight bulkheads forward and aft of the collision. Pumps were activated to eject the water that had flooded into the ship upon impact. The No. 4 hold had taken the brunt of the force of the collision. When it became apparent that the USS *NAPA* had gained stability and would stay afloat, the USS *LOGAN* backed away again.

The high pitch whistle sounded. "NOW HEAR THIS! NOW HEAR THIS! ALL CLEAR! ALL CLEAR! SECURE THE WOUNDED! SECURE THE WOUNDED!"

With a sigh of relief, I grinned at the Marine who had been lying there anticipating his fate without protest. "Ready to go back, fellow?"

"Back to the Grand Palace," he said.

I lifted one end of his stretcher, and another sailor grabbed the other end, and we carried him back down to the infirmary. It took a long while to get all the wounded back, but that was one of the more grateful tasks I'd ever done! Getting them back there meant the powers-that-be were confident that the USS *NAPA*, although definitely damaged, was not going to sink!

By then, dawn had edged over the sky. We proceeded at one-third speed to the transport area in company with the USS *ESTES* and a screen of seven escort vessels. There were two tugs standing by to render assistance. The USS *NAPA* was badly damaged and would be easy prey for diving planes. Speed was gradually increased to eight knots, and as the patched bulkheads held, the speed was again increased.

The call to muster sounded. After a hurried roll call, for we were still in enemy waters, there were six absentees, presumably gone overboard during the collision. I thought we were lucky in all of that confusion to only have lost six. Some of the names were familiar; others were not, but they were all shipmates and I felt sadness for them. There was little hope they could survive if they had gone overboard in the darkness.

I reported to my station in the radio shack. A ship-to-ship message came, and our casualty list just increased. Lt. (jg) Ritchie, Boat Group Officer, had been killed as a result of wounds received on Iwo Jima. He was in command of the 8th Wave and led them to a successful landing on Blue Beach. He was hit by a burst of machine gun fire. He would be buried at sea, from the USS *NEWBERRY*, the ship to which he had been taken by an evacuating amtrack and on which he had died that day.

Even though our ship was heavily damaged, we stayed off-shore at Iwo Jima and unloaded the rest of the equipment designated for that campaign and received more casualties. A hole in its side wasn't going to stop the USS *NAPA* from performing its duties!

At 1630, more than twelve hours after the ramming, the six who had become separated from the ship during the collision came back on board. They had been knocked overboard and were picked up by another vessel. There was no time to celebrate their safe return.

Upon inspection, it was discovered a hole approximately two-feet wide from the main deck down to an unknown depth between frames 93 and 105 was in our ship as a result of the collision. What was destroyed was the electrical workshop, dental office, and sickbay stores. The aft-end of the walindavid (which is what picks up the boats out of the water and secures them on deck) was destroyed. The refrigeration units were torn up, which destroyed all fresh fruits, vegetables and meats. The king post was cracked at the main deck and torn loose at the foundation. The number four hold was flooded to within five feet of the second deck. Two 20mm guns on the port side were out of commission. The fuel tanks were ruptured and fuel oil was on all water in the compartments. The news was not good. The USS *NAPA* was now an injured ship still facing the possibility of another breach in the patched bulkhead.

In the radio room, I sent a coded dispatch to Command: "USS *NAPA* rammed by friendly. Not sunk. Still maneuverable. On way to Guam for repairs."

At midnight, we were lying to in the transport area for the night. We were a severely crippled ship, at the mercy of any kamikaze or enemy shells lobbed from the island. Thankfully, it was, in retrospect, a "quiet" night. We were effectively out of the Iwo Jima conflict. When dawn lit the sky, the USS *NAPA* limped to Guam for repairs.

It was nothing short of a miracle that the USS *NAPA* made it to Guam under her own power. Until we actually pulled into dock at Guam, there was the very real possibility that the USS *NAPA* might yet go down. Although at the time most of us didn't realize how severely the ship was damaged, there was still that uneasiness, that foreboding that the hole in our ship would sink the USS *NAPA*. We didn't voice it. We were all too busy to dwell on that possibility.

At sea, away from Iwo Jima, on 25 February, the crew was called to muster. With the sound of a bugle coming over the innercom, the Chaplain stood before us as an honor guard brought a simple, flag-draped casket on deck. As we all stood at attention, the Chaplain read a brief ceremony and our dead, killed in action on Iwo Jima, slipped over the side of the ship to be buried at sea. The flag that had covered his coffin would be sent to his survivors. It was a somber moment, and reminded me of my own mortality.

CHAPTER 15
Guam

On 28 February 1945, the USS *NAPA* pulled into Guam in the Mariana Islands. There was a seemingly endless stream of ambulances on the dock, waiting for our over 300 wounded Marines. For us, and for our casualties—Marines and sailors alike—those ambulances were a welcomed sight. It was almost as though the ship herself breathed a sigh of relief. Just a week before, those same wounded faced the very real possibility of being tossed over the side of the ship to await a very uncertain fate in the churning waters off Iwo Jima.

After the more serious of the wounded were taken from the USS *NAPA* into the waiting ambulances, the ambulatory wounded began to leave the ship. I watched from the deck as these Marines and sailors, some of them our own, wounded on Iwo Jima, made their way down the ramp to the ambulances. They crowded down to the dock, all brave men who had just walked through the fire of Iwo Jima. It gave me somber pause to think that, as numerous as these were, they were only a small percentage of those who were wounded in the bloodiest battle of the Pacific.

After all of the casualties had disembarked, it was down to the business of repairing the USS *NAPA*. When the dockmaster inspected our damage, he refused to let the ship return to sea before complete repairs were made. It was by sheer luck and the sturdiness of the king post that the USS *NAPA* remained afloat after the collision. If the ramming had been a little more severe, or straight on, the

ship would have been cut into two pieces. The repair master marveled, "Nothing short of divine intervention that this ship made it
here under its own power. Should be shark bait by now."

Having lived through that harrowing night, I couldn't agree
with him more!

First things first! Before the repairs could be attempted, the
damaged areas had to be cleared of debris . . . and spoiled, stinking
food from the refrigeration units.

The spoiled and decaying meat from the refrigerators had to
be removed. A detail was chosen for this thankless task. I was one
of the lucky ones, handpicked to go into the damaged No. 4 hold.
"Geurin," the Lieutenant said, "no need for radio communication
here. Make yourself useful. Get yourself rigged and get down
there."

I donned my swim trunks and, at first, figured this would be a
lark, an adventure. Then, I was lowered, tethered by ropes, into the
standing, stagnant, stinking water down in the hold to the refrigeration area. It was like being submerged in a pail of garbage. Floating
in that mess were rotting meat and the leaking fuel oil. I had to pick
up the oily, grimy stinking meat with my bare hands and toss it into
nets which had been lowered with me for that purpose. Someone
on deck hung over the hole, holding onto the rope on which the
net was attached. When I yanked on the rope, after the net was full,
it was hauled up, and the water from the net gushed all over me.

There were several of us down there, but no time for chatting or
joking. It wasn't the adventure I thought it would be! I tied a scarf
over my nose and mouth, but it did nothing to dilute the smell of
the combination of rotting meat, vegetables, and fuel oil. The meat
was emptied from the remnants of the refrigerator and destroyed.
I was left in the hold to unload the refrigerator until it was completely empty, and to grab the garbage floating in the stagnated
water. It took all day to finish the job. There is no way to accurately
describe the smell and feel of that job! The slime, stink, squish of
the spoiled food attacked all the senses, and it was beyond anything I had ever experienced, or would want to again. The closest
I can come to conveying what it was like would be if you were
to dig into a landfill, and even then it's not quite as odorous or as
obnoxious as that hold. I can tell you that, over six decades later, I
can recall that smell as though it were yesterday. By the time I was

finished, I showered and scrubbed my skin until it was raw to get rid of the stink. Those swim trunks I wore down there got thrown in the trash. There was no way to get the smell out of them or get them clean enough to use ever again.

While we were in Guam, we were allowed to go to the Red Cross hospital to look for any of our buddies we could find who had been wounded and evacuated there. All I remember of that trip are the rows and rows of wounded Marines. The nurses of Guam were overworked, yet they always had a smile and a gentle touch for the wounded. It was amazing to watch their professionalism and personal care.

On 5 March 1945, the USS *NAPA* was in the floating drydock. The pumps began pumping the water out of the damaged hold. As the water level fell, the extent of the collision damage was revealed. The gash in the USS *NAPA*'s side extended well below the turn of the bilges. It was clear that only the presence of the king post prevented the intruder's bow from shearing clean into the keel of the USS *NAPA*.

By 13 March, the USS *NAPA* was moved from the floating drydock to the inner harbor for the finishing touches to the repairs.

While the ship was in dry dock undergoing repairs, I was not exactly having shore leave. When the No. 4 hold had been cleaned of the stinking mess left behind, the Lieutenant decided that I had done such a good job I should be rewarded by being chosen for another special detail. Most of the same crew who cleaned the No. 4 hold were also chosen, although there were a few other lucky ones, too. I spent most of my time chipping paint, repainting, and cleaning the ship. This time in dry dock kept the USS *NAPA* and her crew from getting into the conflict at Okinawa, which was scheduled to be our next invasion. That was a mixed blessing — none of us wanted to die on some foreign beach, but all of us had joined the Navy to fight, not to chip paint in dry dock!

There was another side effect of the collision. Since the refrigeration system was damaged, and the food spoiled, our menu underwent a drastic change. The central item on our menu—for breakfast, lunch and dinner—was Spam! Our cooks were creative and mocked up a menu, featuring Spam. Although the menu was to lighten the mood, the truth was that we did eat Spam for every meal, along with stale bread, sliced bologna and powdered eggs and powdered milk—fried Spam, sautéed Spam, cold Spam, Spam ala king, chipped Spam on toast, roasted Spam, barbequed Spam, Spam sandwiches, Spam Fricasee, Spam en casseroe.

On the morning of 19 March, 1945, I was at my station in front of the radio, taking down coded messages when one came in that momentarily stunned me. The USS *FRANKLIN*, an American carrier that had already racked up an impressive score against the Japanese, including strikes at Iwo Jima the year before, had been attacked 50 miles from the mainland of Japan, and seriously damaged. The message said that the USS *FRANKLIN* was listing starboard at a dangerous angle. There was a warning that the USS *FRANKLIN* was possibly lost, and if it didn't sink on its own, it would be abandoned and scuttled.

The potential loss of the USS *FRANKLIN* was startling. The USS *FRANKLIN* carried a crew of about 3,500 and over 100 planes, and had been instrumental in achieving victory in many Pacific island conflicts. The message said planes were still aboard the USS *FRANKLIN* when a Japanese kamikaze pilot, whose plane was loaded with bombs, slammed into its deck, tearing through it. There were devastating explosions below deck, which dominoed, setting off ammunition, bombs and rockets, causing extensive and debilitating damage to the ship and its crew. The first two decks were ablaze in a river of volatile fuel.

I hand-carried the message across the passageway to the Captain, who took it without comment.

The next morning, there was a follow-up message that said the USS *FRANKLIN* had survived, that it was listing badly, but was afloat and on its way to Hawaii under its own power. This message I took with relief, and, like the other, hand-carried it to the Captain. The USS *FRANKLIN* was severely injured; it has lost lover 700 of its crew, but the survivors had acted heroically and saved many others, as well as their ship.

Although it was good news that the USS *FRANKLIN* was on its way to Hawaii on its own, it was still very much vulnerable to lurking enemy ships, planes, or submarines. No one would breathe easily about the fate of the USS *FRANKIN* until, and if, it came safely into port in Hawaii. All American ships between the Pacific Islands and Hawaii were on the lookout for the USS *FRANKLIN* to render assistance if needed. It would be its longest journey, and from a record of 33 knots (less than 50 mph), it was down to a limping 13 knots (less than 15 mph).

On 25 March, the repair work on the USS *NAPA* was finished. That morning, at 0845, we took aboard Japanese prisoners of war (POWs) under the watchful eyes of Marine guards. Some of the prisoners had been captured on Iwo Jima. The POWs were placed in the newly redone number four hold. Armed Marine guards were posted at the top of the ladder inside the hold where the Japanese would have to emerge for bathroom privileges. The Marines had no love for the prisoners.

I watched from the just outside the hatch to the radio room as the POWs filed by below and were systematically searched and then directed down the steel ladder into the bowels of the No. 4 hold. They looked young, hardly more than teens. In spite of their youth, I couldn't feel sorry for them, for they were the same Japanese who fought, and killed, our Marines on Iwo Jima. It was because of them my friend Hornick had died back there.

Garner came out to join me. I said, "They're kids." And shook my head.

Garner laughed. "Hey, Geurin," he said, "you're not exactly an old man yourself."

Funny, but I didn't think of myself as a young kid anymore, but he was right. Those POWs going down the ladder into the hold were about the same as I was. I was only 20 then, still not legal to buy a beer in the States, but an old man in war years.

Most of the Marine guards were the ambulatory wounded who were being evacuated. Their injuries had been inflicted by either the POWs they now guarded or their comrades. It was a tense situation. I spoke with the Marines and I knew it would take very little to provoke them into dispensing with any prisoner who challenged them in any way. Thankfully, that did not happen. This was still war, and we were still in the Pacific, still in danger from Japanese attack, even though Guam was a secure base. Looking back through the years, it may be difficult for some to understand the feelings of the time, but being there, seeing the enemy face-to-face put a different perspective on it. These prisoners may have been rendered harmless now, but it wasn't all that long ago that their one objective was to kill as many of us as possible. It still was their objective.

Although some of the sailors did converse with the POWs, most of them watched them with malice cultivated by the battles that they had just fought, and which cost the lives of many sailors and Marines. After all, we had just come from Iwo Jima, buried our dead at sea, and off-loaded over 300 wounded Marines and sailors at Guam! The Japanese soldiers were not exactly our friends! I'm sure they viewed us with malice, too. After all, in their culture it was a disgrace to be a captive, not to have had the honor of dying for their Emperor and country. The duty of any prisoner of war was to escape or cause havoc or chaos among the enemy, and so they were watched very, very carefully. Any attempt on their part to cause harm to themselves or others would be taken care of permanently.

That day was also Palm Sunday. The three leading faiths—Catholic, Protestant and Jewish—held services on board the USS *NAPA*. In time of war, there is standing room only at any faith service, and that time was no different. Those who chose to worship differently were given their time and privacy to do just that. I tell you, there are very few—if any—atheists in a war zone! I attended the Protestant services.

At 1100 hours on 25 March 1945, we took aboard ambulatory and stretcher patients, both Marines and Navy, for transportation to Pearl Harbor or the States.

"What you think, Estrada? Pearl or the States?"

He shrugged. "Either one's better than here," he answered.

How true that was! While we were hoping to be sent to the States, we did not know where we were headed when we left Guam. However almost anywhere would be better than sitting in the Pacific among the islands not far from the Japanese homeland. I had learned long ago that with the Navy—especially in a wartime situation—there were no guarantees. Even a Radioman accepting and decoding messages wasn't told where the ship was headed.

We made a friendly bet as to where we'd end up. Having the wounded Marines aboard ensured that we would be heading to one of those places, either a safe harbor. It would be good to be on friendly soil again.

On Tuesday, 27 March, I was just off duty and was walking down the narrow aisle way toward the galley when I had to squeeze around two guys rolling dice against the bulkhead. They scooped them up and stood when I neared them. When they saw I was not an officer, the one with the dice tossed the dice up in the air and caught them. "Hey, Geurin," he said, "want to throw dice with us?"

I shrugged. I had nowhere else to be that couldn't wait. "Sure," I answered.

He looked at his buddy, then at me and said, "For a ten?"

"Why not?"

"Since I have the dice, want me to go first?"

A gamble's a gamble. I'd tossed a few dice here and there. Didn't matter who shot first.

He tossed the dice. A seven! He tossed them again. Another seven! He grinned and held out his hand, and I dug in my pocket and dragged out a ten. I snatched the dice and felt them. Now I couldn't be sure, but what were the chances of throwing two sevens in a row like that? I tossed the dice back to him. They were loaded. No chance otherwise. Need I say those two weren't on my Christmas card list!

CHAPTER 16
Pearl Again

Five days later, still at sea, on 1 April 1945, we celebrated Easter Sunday. At sunrise, Protestant Holy Communion Service was held on the forward deck of the USS *NAPA*. A Catholic service, led by Lieutenant Commander John O. Bracken of Baltimore, Maryland, was given on the forward boat deck. An hour and half later, a general service was broadcast over the ship's public address systems for all the casualties and others of mixed faiths. That Easter Sunday, with no organized church, with family made up only of battle-weary crew and survivors of the island fighting, we had a great deal for which to be thankful. We had survived walking through the fires of Iwo Jima, a collision that was a hair's breadth from fatal, and we were on our way out of the dangerous waters to a friendly port— either Pearl Harbor or Seattle, it didn't matter. Everyone on board that Easter welcomed dawn with fervent prayers. I was thankful just to be alive. A better Easter, there never was!

In spite of the relative serenity of Easter services, we were ever vigilant that we were still in enemy waters. The lookouts, in their battledress including flak jackets and helmets, continued to scan the ocean for any telltale signs of German underwater boats. The antiaircraft gunners stood round-the-clock duty, swiveling their guns to scan all around us. There was a constant surveillance of the skies for Japanese war planes. In the radio room, I monitored ship-to-ship and shore-to-ship communications, as well as scanning for enemy radio waves. It was possible, even then, to pick up a certain amount of chatter and static from German U-Boats. Even though we had an interpreter onboard, it wasn't really necessary to know what they were saying if we were close enough to hear them, they were already too damn close to us! Thankfully, I didn't hear any.

It was during one of those times that we received word we would be going to Pearl Harbor. I still had on my earphones when I gave a thumbs up to Estrada. He clapped me on the back and nodded. He had already heard the same at his station.

On that cruise from Guam to Hawaii from the war zone, there was a spot of levity. A loosely formed, shipboard entertainment company that called themselves the NAPA Nutcrackers put on a performance for the crew and passengers. On the playbill were Scintillating Saso (Domenic Saso, RM/2C from San Francisco, California) on the squeezebox and Truetone Vella (LaMarr Vella, SSMB/3C of Lyman, Wyoming) on the trumpet, who led the company in an unforgettable performance. Big Johnny Johnson gave his immortal rendition of "The Road to Mandalay" to an audience of cheers and whistles. That was followed by Big Stupe Sullivan (Chief George Sullivan of San Francisco) and Barrelgut Krause (Eugene Krause, MoMM/3C, of Hawley, Pennsylvania), with their sterling winning performance of Chocolate Meringue Eating. To foot stomping, more whistles, and calls of "Way to go!" the smiling winners wiped the chocolate from their faces and accepted the accolades of their captive audience. They would never win an Oscar or a Tony Award, but for us on the USS *NAPA*, they were pure gold. They gave us a much-needed laugh in the middle of a rainfire of horror.

In spite of the light heartedness, the war was not forgotten. The wide open ocean between Guam and Hawaii was still dangerous. The lookouts kept their keen eyes and binoculars trained on the both the sea and the sky for any irregularity.

Meanwhile, inside the radio room, I knew we were heading for Hawaii, but any destination past that was yet to be announced. This would be a different type of entrance to the place where the war had started.

On 5 April 1945, we pulled into Pearl Harbor, Hawaii, for the second time. The first time I saw Pearl Harbor, I began to realize the war was a reality. Now, having gone through the battle of Iwo Jima, I no longer had romantic notions about war. It was real. It was here. It killed and mangled, and turned ordinary boys into matured men. I looked at the scars of Pearl Harbor with a new understanding. The cries of those who had died there seemed to echo through the harbor as the USS *NAPA* pulled into Berth B-17.

Someone on deck hollered, "LOOK!"

Most of us had gone topside to watch the entry into Pearl Harbor. I followed the pointing fingers, and there coming into the harbor was a dangerously listing ship. As the USS *NAPA* maneuvered through the harbor to go to its berth, the ship we were watching came into clearer view. It was the USS *FRANKLIN*! The front of its hull was missing. The top deck was mangled to the point it resembled a large slice of Swiss cheese. Its large antiaircraft guns were twisted and blackened. There was a large, gashing hole in its side, the metal of at least two decks peeled back like a tin can under the blade of a can opener. Yet, it steamed into the harbor on its own.

On deck, a single solitary clap became a spontaneous ovation. The crew aboard the USS *FRANKLIN* could not have heard us or seen us, but it didn't matter. The heroic ship had come home from halfway across the ocean, mangled, wounded, crippled, but under its own power!

The USS *FRANKLIN* would undergo emergency repairs in Pearl Harbor and then make its way back to its home base in Brooklyn, New York, completing a journey that should have been impossible. The USS *FRANKLIN* became a courageous symbol of victory at sea.

The wounded and the POWs were off-loaded at Pearl Harbor. On deck, the crew of the USS *NAPA*, along with the ambulatory Marines who had been wounded on Iwo Jima, watched with undisguised malice as the first of the prisoners were brought up from the No. 4 hold. These were still the enemy, and they represented all of the reasons we were here. Some of the prisoners who were also seriously wounded had to be carried off the ship on stretchers. The stretchers were carried by other POWs under the ever watchful eye of armed guards, this time Navy guards from Pearl Harbor.

The American casualties of the island engagements in the Pacific could take their first safe breaths. Watching them go ashore was a replay of Guam. Some were still dazed; some somber, but all were happy to be at Pearl Harbor. It took nearly two hours to transfer all of the wounded, and then the POWs, from the NAPA!

As I watched the POWs being escorted off the ship, I saw an

amazing sight. There was a wounded Marine being helped by a Japanese prisoner. They walked down the ramp together, with the POW very obviously assisting the American. I didn't feel there was a camaraderie there or even a trust. It seemed a matter of expediency, and putting aside for the moment the reasons that caused them both to be there.

I was leaning on the top railing of the deck, arms on the rail, hands clasped, as that extraordinary scene played out below me. I nudged a sailor next to me. "Did you see that?" I asked, not taking my eyes from the unusual duo.

"Yeah," he answered, "I don't think I could have done that. How about you?"

I thought about it. Would I have allowed myself to be assisted by one of the enemy? Under those circumstances, on our ship? I still don't know, and I'm glad I wasn't tested in that way. The two disappeared into the now crowded dock, and I lost track of them. I knew they had to have been separated when they reached the shore.

When all of the Marines and prisoners had disembarked, the USS *NAPA* commenced taking on supplies. It became obvious that the USS *NAPA* and her crew were far from being out of the war.

Three days later, on 8 April 1945, the USS *NAPA* had a change of command. Captain F. Kent Loomis, USN, from Vallejo, California, reported aboard to take charge of the USS *NAPA*.

Within another two days, the USS *NAPA* was again in dry dock, this time in Pearl Harbor, so that her hull, which had undergone emergency repairs in Guam, could be restored to its original specifications. There was still a war going on, and the USS *NAPA* and its crew had not yet been mustered out of it. The ship needed to be completely seaworthy.

While in dry dock, with most of the crew aboard, the warning whistle sounded, loudly, shrilly: "NOW HEAR THIS! NOW HEAR THIS! THE CAPTAIN WILL MAKE AN ANNOUNCEMENT. NOW HEAR THIS!"

I remember I was on deck, walking toward the hatch that led down to the galley. I stopped, leaned casually on the railing, to listen. The Captain rarely made an announcement himself, so I was curious, as I am sure all the crew was.

The Captain's voice was unusually somber. "It is my sad duty to tell you that President Franklin Roosevelt died on 12 April."

I know I was not the only one who caught his breath in disbelief.

The Captain paused, then continued, "Harris S. Truman has been sworn in as President of the United States. That is all."

President Roosevelt dead? It was unbelievable. He had been first elected President in 1932, and relected all through the war. He had held that office longer than any. It was impossible to think of the United States being led by anyone else during this time of war. Harry S. Truman was little known then, and I could not imagine his taking the reins from President Roosevelt and leading us to a successful conclusion to this war.

It wasn't long before I learned the sketchy details, that President Roosevelt had been in failing health and died of cerebral hemorrhage at his home in Warm Springs, Georgia. The fact he had been in deteriorating health had been kept from not only the press and the American public, but also most of his advisors, as well as his Vice President Harry S. Truman. It was reasoned that during a time of war, especially one that covered so many fronts, it was ill-advisable to let the enemies of the United States know that our president was in poor health. Even the fact that the president was wheelchair-bound as a result of polio was not widely known.

The flag was lowered to half-mast, both on the ship and on the base at Pearl Harbor.

Being in dry dock did not stop the normal activities aboard the USS *NAPA*. It also encouraged free time athletics as a way to pass the time until the ship was at sea again. One of those sports was the boxing matches. A boxing ring was set up on deck, and was fairly authentic in size and looks, considering we used what we had at hand to construct it. Three tiers of rope were strung around four posts that were anchored by sheer tension. The boxing mat itself was just canvas spread over boards.

The boxers had genuine boxing gloves, courtesy of Stores, and wore their swim trunks or casual shorts and athletic shoes in the ring. There were some fairly good boxers aboard the USS *NAPA*. The captive audience stood or sat anywhere we could, even on the barrels of the big guns. Friendly bets were always taken. There were no holds barred, and the boxing became fairly serious, but never bloody. They put on quite a show.

After one such match was over, that I watched from a perch

near the radio room, S1C Alvin Stanfield, originally from San Pedro, California, said, "I'd beat your ass in the ring anytime, Geurin." He challenged.

"You couldn't beat a girl," I sniped back.

"Talk's cheap," he said.

I had boxed some in high school, but it was not my sport of choice. I really thought he was kidding. "Back your talk, then," I said.

"Damn right," he answered.

I nodded toward the ring where another two boxers had entered and were being talked through the rules. "Ring's busy," I pointed out.

"This can be private, Geurin."

He was serious, and even though I knew him slightly, we had nothing in common and there was no reason for the animosity he was exhibiting. "You serious?"

"What's matter, Geurin?" he asked. "You afraid to take me on?"

I shook my head. "You're on. Where and when?"

"No. 4 hold," he said. "Now."

I was off-duty and it seemed that this issue would be pushed until it was settled. I nodded. He left me standing there. I did notice that another fellow went with him. I wasn't afraid of being ganged upon, for that wasn't the way of the US Navy. I just wondered why he had challenged me to start with.

In the past, mostly in my civilian life, I had had shorter guys egg me on, spoiling for a fight. Mostly, I avoided it because, really, what was the point? Stanfield was a couple inches shorter than me and I vaguely wondered if that was part of the reason for the hostility he showed. Regardless of the reason, I was committed.

I went down to the No. 4 hold. Stanfield and his friend were already down there, with the boxing gloves. He handed me a pair, and I put them on, tugging the second one on with my teeth. His friend, to whom he did not introduce me, laced my gloves, then laced Stanfield's.

The second man said, "Ready?" and looked first to Stanfield then to me. We both nodded. "No punching below the belt or to the throat." We both nodded. "Step back," he said, and Stanfield and I moved away from each other. "Proceed," we were told.

Stanfield came out with a good boxer's stance, dancing on the balls of his feet, one gloved hand about nose-high and the other just in front of his chest. He did a little dance step and made short punching motions with his hands.

I smiled. I stretched out one arm and held him away from me. There was no way he could reach me, for his arms were much shorter than mine. He pressed against my outstretched arm, tried to duck under it, but I kept up the pressure. He was held fast like a dog on too short a leash.

"Let me at him! Let me at him!" his friend shouted.

Stanfield tried again to come in closer, but I had the advantage of a long reach and held him away from my body.

"Let me at him! Let me at him!" his friend shouted again.

I turned Stanfield loose and skipped back out of reach of his punch. He stopped, looked at me, then his friend, and nodded.

I stayed where I was while his friend helped him out of his gloves. Then Stanfield pushed the gloves on the hands of his friend and laced them. He didn't go through the same cautionary routine, but moved aside and his friend charged.

They were the same size. He tried to avoid my reach, but it wasn't possible. I moved in and stretched out my arm and pushed against his chest. He punched the air, hitting this way and that, until it became apparent that he, like Stanfield, simply did not have the reach to get to me.

"Enough?" I asked.

He bit his lower lip and nodded. "Enough," he conceded.

So it was without a blow landed by any of us the impromptu boxing match was over. Stanfield and I never became friends, but neither did we hold animosity toward each other.

When I returned topside, the official boxing was still going strong. I no longer had the appetite to watch, so I went on into the radio room.

One advantage of being a Radioman was that I heard information before the general population did. Often times, I could not disclose it because many of the communiqués were secret and coded, and meant only for Command. It was good, though, to know where we were going after Pearl Harbor. I said to Garner, "I'll bet we'll be heading back to the islands." There was no doubt in my mind we would be going back into the heart of the war that was still going on out there in the Pacific.

"You're on," he said and gave my hand an open slap, a gentleman's agreement.

A short time later, on 30 April, word came that Adolf Hitler, dictatorial leader of Germany, had committed suicide. The war with Germany was anticipated to be over within days. Even though we were in the Pacific, and not in Europe, this came as great news. Germany had been steadily losing ground against the Americans and the Allied Forces, and it was only a matter of time before the complete collapse of the German forces.

Two days later came the news that didn't come as a surprise, considering that Hitler was dead. On 30 April 1945, Germany unconditionally surrendered. The war in Europe was over. Italy had surrendered the year before on 3 September 1943 and its brutal dictator, Benito Mussolini was dead, executed in a public square by his own people. Now Hitler was dead and Germany surrendered, bringing the war in Europe to an end after six catastrophic years. Europe had been devastated by the war, and many of its cities destroyed, countless civilians among the dead, but their war was over. The only one left of the Tripartite Treaty was Japan, which had not only not given up, but was fighting stronger and more furiously than ever. On 7 May 1945, the official surrender of the German armies took place.

The bet with Garner was one bet I was glad to lose. In mid-May we received word our next destination would be Seattle, Washington. The repairs had to be finished first, and they were nearly done at the time of receiving the orders. It would turn out that we both were right, but first Seattle.

Late in May, the repairs were completed, supplies loaded, and the USS *NAPA* was ready for her next assignment.

CHAPTER 17
Seattle

On 23 May 1945, we cast off for Seattle, Washington, to take on troops to return to the Pacific. The journey from Hawaii to Washington took seven days. Even though we were in friendly waters, we did not let down our guard. Ships had been continually sunk off the coast by German U-boats. The expanse of ocean from Hawaii to the United States coastline was no place to relax our guard. As with the journey from Guam, I stayed on duty in the radio room to monitor radio traffic. The closer the USS *NAPA* came to shore, the more chatter on the radio, and the more important it became to identity it to be sure no U-boats were prowling nearby.

One week later, on 30 May, we anchored one mile from Ferry Landing in Seattle. Eight months in the Pacific made that city much more attractive than it was on our shake down cruise the previous summer. Eight months. It seemed like another lifetime ago.

We were given liberty. Jack A. Kapp, a Signalman 3rd Class and a good friend, who was originally from Washington D.C., and I went into Seattle. It was a time to forget, for a while, what we had seen—and, most probably, would again see—in the Pacific. Liberty for sailors was as necessary as breathing! When all we had was sea for weeks and months at a time, interspersed with the fire of the enemy, dodging the kamikaze, and fighting to stay afloat, a few weeks in a friendly stateside port seemed heaven-sent.

There was a USO in Seattle that provided doughnuts, hot coffee, cold soft drinks, ballroom music, and, most importantly, female company. The USOs were relatively new at that time, having been formed in 1941 in response to a challenge by President Franklin D. Roosevelt. The President recognized that military personnel, especially those stationed far from home, needed a friendly and safe place to go for rest and recreation. He challenged six private organizations—the Young Men's Christian Association (YMCA), Young Women's Christian Association (YWCA), National Catholic Community Service (NCCS), National Jewish Welfare Board, the Traveler's Aid Association and the Salvation Army to come up with a unified organization that would handle on-leave morale needs for members of the Armed Forces, without prejudice, without any private, public, religious, or governmental agenda. The result was the United Service Organization, a strictly volunteer group that provided much-needed islands of peace where a soldier, sailor or Marine could forget the war for a day or a night. The USO served not only hot coffee and cold drinks, but entertainment in the form of free dances.

At the Seattle USO, the girls were pretty and the music was hot. Cold soft drinks were a nickel for civilians, and free for those in the uniform of the Armed Services. There were doughnuts and fresh baked apple pie and homemade cakes. There was hot coffee served with a warm smile. It was always noisy and warm at a USO. The music usually was from on-stage bands, and it didn't matter if they were local or well-known groups. It was perfect for a sailor on leave.

I loved to dance, and there was a particular girl at the USO who made a good partner. There were rules there that the girls didn't leave with any of us, and that was all right with me. That girl, Helen Kilmer, was there every night and she would leave others on the dance floor just to dance with me. After a while, I found myself eager to go there just to see Helen. It didn't matter that she and I were breaking the rules by seeing each other exclusively.

One night, we stepped outside so I could smoke. The night was cool, the sky clear, and the war far, far away. Helen and I sat on a wooden bench and listened to the fog horns and the occasional car horn from the street in front of the USO. I felt like life would be perfect if Helen was by my side always. Romantic nights far from home have been known to do that to a fellow.

Helen lay her hand in mine, and I squeezed it. I couldn't think of anything more perfect. I stood, drew her up with me, and we danced there in the moonlight outside the USO. I sang "Sentimental Journey" in her ear, and although I couldn't carry a tune and would never be mistaken for Mario Lanza, Helen didn't seem to mind at all! When our slow dance finished, I took both her hands and told her, "I'm crazy about you, Helen."

"I'm crazy about you, too, Arvy," she said.

We kissed there in the moonlight and walked slowly back to the USO. Before we got inside, I told Helen, "I'm going back over there, and I don't know for how long."

"I know," she said, "it doesn't matter."

"Wait for me?" I looked down at her and at that moment she was all that was important in the whole world. It was the music, the night, and the pretty girl.

She smiled, squeezed my hand again, nodded. She reminded me, "I'll probably be back in Chicago when you come home. Here." She pressed a piece of paper in my hand, on which was written her telephone number in Chicago.

I tucked it in my pocket, and we went back inside. The music was playing loudly, voices of men and women overlapped each other, glasses clinked together, waves of laughter rolled over all else, and the hearty smell of fresh doughnuts wafted in the air. But for me, for that night and that space of time, all that was important was the girl by my side.

For the remainder of the time that I was in Seattle, I saw Helen every night. We spent more time outside the USO than inside it. We made promises to each other, and plans of what to do after the war. "One day," I told her, "I'm going to open my own bakery. It'll be the finest in all of California."

Helen stroked my face gently and said, "I'll work right by your side. We'll have the best bakery in the world."

Then came the day when I knew the USS *NAPA* would be on its way back into the heat of battle. Helen promised she would write regularly.

It was a fine June day when I found out we would be leaving Seattle. When I went back out to sea into the Pacific, I knew I would be re-entering the war—we all did. There was no telling where I'd be going or for how long. I'd survived walking through the fire of

Iwo Jima, and it did no good to think about if I'd survive another battle. That's what I was there for—to go to battle as long as it was necessary.

CHAPTER 18
Army on Board

By mid-June, the USS *NAPA* was ready for duty. On 16 June 1945, cargo was loaded in preparation of our journey back to the war. Three days later, Army troops came aboard. Looking down over the ship, I was amazed to see the long line of troop transports spilling out hundreds of Army troops, readying to come aboard the USS *NAPA*. These would be taken only as far as Hawaii, but I didn't see how even the USS *NAPA* could hold so many! We managed, through, but it still astounds me how we did. Looking over the ship at all those soldiers, and I wondered how much of the United States was represented there. There had to be soldiers from every state.

I was one of the sailors lining the ship's rail, looking over the side at the seemingly endless line of Army personnel swarming all over the dock. They kept coming and coming as though some human floodgate had been opened and spilled them out.

The soldiers checked in on deck and then were sent down to the third and fourth decks. There, Stores dispensed folded, canvas cots and directed the troops to an area usually reserved for cargo. The Army men each chose a spot, and dumped their duffle bags and cot there. It was crowded; very little walking space between the temporary bunks. But these were US Army soldiers, some of whom had already experienced sleeping in far worse conditions than a crowded hold of a ship. For the rest, those who were fresh-faced and eager to go to war, the hold of a transport vessel would be considered hands-on training for what was to come on those islands in the Pacific.

There wasn't much interaction between the US Army and the US Navy. We were a taxi for them, nothing more, and we didn't see that becoming fast friends with any of them would benefit any of us.

One hot day, halfway between Seattle and Hawaii, the sea was particularly rough. There was a small group of soldiers who spent their day sunning themselves on the top deck. When it was chow time, they didn't bother putting on their shirts, but went below to the galley all sweaty and smelly. None of us particularly liked being anywhere near them, but our chow line had an order to it and once in it, there was no way to exit it until our metal trays were full. That is unless one wanted to go back to the end of the line and start all over. That wasn't a very feasible option considering how many people were aboard.

That day, with the sea churning, the ship rolled with each forward wave, and then settled when the wave passed. The back-and-forth rolling motion was second nature to sailors, but not to our Army guests. I smiled when I saw any of them clutching the side of the ship, or stumbling from table to table in the galley in order to keep their balance.

When we were done with our food, it was required that the metal tray with its leavings be taken to the end of the chow hall where there were large, metal barrels in which to put whatever was left on the tray. Then the emptied tray went on a conveyor belt that took it into the back of the kitchen to be washed.

While I waited for my turn to dunk my metal tray in the barrel, I noticed that the barrel wasn't fastened down. Apparently, after it was last emptied, someone had forgotten to strap it back in place. It moved with every roll of the ship. It would scoot forward a bit, and then roll backwards when the ship settled. Four of those obnoxious, shirtless, sweaty, Army men happened to be near me while I waited in line. I watched the barrel move ever so slightly forward, and then back again. It got closer with each forward movement.

I went to the barrels and dipped my tray into one with more force than was necessary for the leftovers to fall into the barrel. I don't know to this day if my extra force was the deciding factor or the coinciding high bump caused by the uneasy sea. But just as I walked away toward the conveyor belt, the barrel tipped. Out spewed the garbage from scores of trays!

Now I admit I could have hollered, "Watch out!" but because I had been standing behind those half naked sweaty bodies enough to gag on the smell, I deliberately chose not to warn them. I made sure that my feet were well out of the way as the messy, stinking, liquid-fueled garbage splashed all over the hapless Army men.

I suppressed my laughter until I was well clear the mess hall.

One day, on the way down to the latrine, I heard an aggrieved shout, followed by an obscenity. Curious, I hastened my pace. There, kneeling on the floor in front of the long trench-like latrine were two of the Army's finest, their heads deep in the trough of the toilet. It wasn't unusual to see those not in the Navy lose it, especially when the sea got rough. To see someone attempt to lean over the railing topside was common when ferrying non-Navy personnel. This time, though, the two soldiers were kneeling in front of the latrine already in use! They didn't particularly care who was using it or why. I watched as they were pushed away. They just crawled right back, grabbed the rim of the latrine and ducked their heads over it and heaved again, until there was nothing but dry, hacking cough left. They didn't even care that some of the sailors were not particularly careful where they aimed.

I was at my station in the radio room ten days later when we moored at Pier 39-D in Hawaii on 26 June. I stepped outside the hatch and watched the disembarking. For that mass of Army personnel we had taken on in Seattle, it only took one hour to off-load them. I wonder, in retrospect, if they were as anxious to get off the ship as we were to have them off?

An hour later, we off-loaded cargo. Then came the announcement for which we had all been waiting. The shrill whistle was followed by: "NOW HEAR THIS! NOW HEAR THIS! SHORE LEAVE FOR ALL OFF-DUTY PERSONNEL COMMENCES NOW. NOW HEAR THIS! NOW HEAR THIS!" No one had to tell a sailor that twice! I scrambled off that ship much faster than the soldiers before me. I was in one of the first groups to get ashore.

I knew that Hawaii was only a jumping off point. I had a brief shore leave in Wakiki. It was no secret that the USS *NAPA* would be heading back into the war. There were no dances, no island cruises, no vacation in paradise. Wakiki in war time was not the island of

travel brochures. It was a beach that was used for practiced invasions, so there were no tourists gathering for luaus, no fancy hotels
looming over the sapphire blue water, no girls in bikinis. But it was
a good place for a sailor to get beer and roam around on friendly
soil before going back to the serious business of war. There was
also plenty of beer to be had. A sailor was not far from his beer. As
one of my friends remarked, "A sailor's nose is as fine tuned to finding beer as a bloodhound's is to finding a rabbit."

Although Pearl Harbor was still a wounded base, and all non-
essential personnel had been evacuated shortly after the bombing,
the island had recovered greatly since then. There would be no for-
getting of that terrible Sunday, but life did go on. Businesses rebuilt,
and hotels reopened, but there was a wariness that wasn't there
before 7 December 1941.

For the most part, I stayed with Jack Kapp. We had been encouraged to stay in pairs. It wasn't unusual for a sailor, who was too
much into his drinks, to fall victim to a mugging. To tell the truth,
even going in pairs didn't stop that, because if one imbibed sailor
was too out of it to ward off a mugger, two were just that much
more inept, which gave the mugger twice the pickings.

For Kapp and me, we were just smart enough to enjoy ourselves
without getting to the point that we wouldn't remember if we enjoyed ourselves or not!

There was a hotel in Honolulu that catered to servicemen. They
welcomed us with open arms, served our every need. In their bars
and restaurants, they sold beer and whiskey, and encouraged me to
buy more and more. It was like being in the lap of luxury compared
to being on the ship. It didn't take much to be convinced to stay to
the limit of my leave.

What one has to remember in retrospect is, in spite of the ste-
reotype of "drunken sailor," what we really were, were young men
in the brief oasis between war zones, who knew that, once our
ship set out to sea again, there was a very real chance any one of
us might not return with it. It wasn't something on which we dwelt
. . . I know I didn't . . . but it was something that was there at the
edge of our minds. When given the chance to forget all of that for a
while, we took it, and took it with the relish of enjoying life for the
moment. That stay at the oasis was far too brief.

On 28 June 1945, the USS *NAPA* received cargo for the troops fighting in the islands. Most of the cargo was ammunition and other battle supplies, but some of it was life stores (such as, uniforms, blankets, food, shovels, rations). We also received supplies and provisions from the US Naval Supply Depot for feeding the crew and troops on board. It was amazing just how much food the ship's complement and passengers required: 20,000 pounds of wheat; 5,000 pounds of sugar; 1,200 pounds of pumpkins; 2,400 pounds of rice; 2,465 pounds of sweet potatoes; 350 gallons of salad oil; 1,000 pounds of beans; 192 pounds of dry soups; 2,948 pounds of peas; 706 pounds of dry cereal; 1,500 pounds of powdered milk; 1,170 pounds of turnips; 1,980 pounds of lunch meats; 1,847 pounds of bacon; 2,120 pounds of pork loins, 3,000 dozen eggs, 1,980 pounds of butter, 800 pounds of beets, and 300 pounds of yeast. I tell this because it very much illustrates the impact of what was destroyed off the coast of Iwo Jima when the USS *NAPA* was rammed by the USS *LOGAN*. It also illustrates what I, along with other lucky fellows, had to remove from the No. 4 hold. Only then, those items certainly weren't fresh and new, but spoiled and mixed with motor oil, and decayed and soaked by dirty and stinking water for days.

When the cargo was fully loaded and stored in the holds, the troops came aboard. This time, they were Marines again. They, just as the Army men before them, took up a lot of space in the holds, and they, too, slept where they could on the canvas bunks. By then, I was getting used to the overcrowded conditions on board the USS *NAPA*. After all, one of our missions was as a troop transport.

All of this time, Helen Kilmer and I had been writing rather sporadically to each other. I wasn't a great letter writer, and I knew that Helen was busy with her own life and the USO. Getting mail wasn't easy aboard ship in the middle of war. The supply ships with the mail bags caught up eventually, and sometimes there would be a thick pack of letters; other times, there'd be one or two. My letters home to her would also have been received on an irregular basis. I had had time to think about those idyllic nights in Seattle, and the magic of them had faded with the passing of both miles and time. One night I sat on my bunk, my legs dangling over the sides, and wrote: "Dear Helen, soon, the NAPA is setting sail again. I don't know the destination, but I do know it will be somewhere in a war zone. I don't know where or for how long. We made a lot of promises, and I meant them. But I don't think it's fair to hold you to them. It's foolish to think that you should wait around for me, when I don't when or if I will be back. If and when I do return, we could meet again. If you feel the same way then as that night in Seattle, then will be the time to talk about our future." I read it over twice, changed a word here and there, essentially left it the same, then signed it, "With love, Arvy." Before I could change my mind, I took the letter down to the mail room for dispatch. I don't know if she ever received the letter, or if she did and it must have been taken seriously, because she never again wrote to me. In retrospect, I think I was a little disappointed that she accepted my decision so easily, and didn't put up a protest at all. Maybe she had felt the same way and was just too kind to tell me.

CHAPTER 19
Eniwetok and Ulithi

On 29 June 1945, the USS *NAPA* prepared to get underway. In the radio room, I was alerted to our destination: Eniwetok Atoll in the Marshall Islands in the Pacific. We were heading to back into the war! That did not come as a surprise, if only because of the Marines aboard. The USS *NAPA* joined the convoy of American fighting ships. We moved out of port, and we zigzagged away from the Hawaiian Islands.

As we sped toward the Marshall Islands, the USS *NAPA* crossed the International Date Line and 1 July 1945 was lost forever in that time warp. That was the second time I lost a day.

"At this rate," I told Garner, "if I lose many more days, I'll be in the future without having a past."

Garner laughed with me. This losing a day by merely crossing an invisible line did seem odd to both of us.

That Fourth of July came and went. In spite of its awesome meaning, and the amplified patriotism aboard ship, it was just another day at sea. There were no fireworks, no flag waving, no parades, and no backyard barbeques. All the ways I used to take for granted in celebrating this holiday meant nothing out on the ocean in the midst of a war. I spent my free time washing out my socks. In retrospect, one might think it strange that a shipload of military men would ignore the Fourth. The fireworks we had all seen made on Iwo Jima made the symbolic fireworks in cities across the States seem tame. The words to the National Anthem were never truer than to a man was at war. *"Rockets red glare ... the bombs bursting in air, gave proof through the night that our flag was still there."* The flapping flag rising from the mast of the USS *NAPA*, flying in the wind, was proof enough for us. That flag would never dip in surrender; never be lowered from its mast, as long as we were out there fighting for it.

On 7 July, we anchored at Eniwetok. General quarters sounded! The USS *NAPA* began a serious round of war games. The big guns fired ... **KABOOM! KABOOM!** Big, black puffs of smoke blew back over the deck as the guns fired again and again! Battle stations blasted over the ship's speakers. I was already on my way to the radio room. I clapped on my earphones as I hurried to my station at the far end of the radio shack. I received and sent coded transmissions to the other ships in the convoy. The war games were very practiced, very choreographed, in order that when we entered the real battles, we were all prepared. There wouldn't be time in the midst of battle to decide who did what and when.

The orders came for the beach party to prepare for debarkation. I left the radio room, grabbed my helmet, flak jacket, and my section of the radio and dashed for the boats. I climbed down the rope ladder into the landing craft and crowded in with the others. The coxswain pushed off and we made a landing on Eniwetok. This practice landing was far more realistic than the first ones I had participated in, in both Oceanside and Maui.

We hit the beach at Eniwetok, scrambled to get our radio assembled, and sent messages back to the USS *NAPA*. It went smoothly this time, much better than what happened on Iwo Jima. All the while, the ships fired their big guns toward the beach over our heads to simulate authentic engagements with the enemy. When given the all clear, we then darted back to the landing craft and were taken back to the USS *NAPA*.

This drill went on and on, with the goal that it would become remote, and as much as part of us as breathing. The following day, the drills were repeated, with more intensity. By 14 July, we had hit the beach so often I was dreaming about it. That morning, the USS *NAPA* pulled away from Eniwetok and headed out to sea again. The big guns were fired again and again, until their boom continued to echo even after the drill was done.

My regular shift was over, but during drills, no one was ever completely off-duty. I kept my gear nearby.

We sailed toward Ulithi Atoll in the western sector of the Caroline Islands Chain, approximately 360 miles from Guam, and in direct line to Tokyo, Japan. The previous year, in September 1944, the US Navy took control of the Ulithi Atoll. The islands had been abandoned by the Japanese as the Americans advanced across the Pacific.

The Ulithi Islands became the largest US Naval Base in the area. The Navy soon amassed a large contingent of ships, including battleships, aircraft carriers, destroyers, cruisers and other support ships at Ulithi. It was a perfect strategic location from which to launch an all-out assault on Japan.

All of this time that we were out to sea, we continued practicing with the big guns firing, and all of us to battle stations. Every morning, we had general quarters at 0500. We were constantly on guard, in full battle gear, constantly aware that we were still at war and that at any moment we could be blown out of the water. After having been in one battle, we now knew the realities of war.

In the morning of 18 July, I happened to be out on deck for a smoke when the Islands of Zohnoiiyoru Bank were spotted. I quickly put out my cigarette, field stripped it, put the remnants in my shirt pocket, and hurried back into the radio room. Four hours later, just as I getting off duty, the USS *NAPA* anchored at Ulithi Atoll in the Caroline Islands.

One of the islands of the Ulithi Atoll was known to the sailors as Mog Mog . . . an island in the Pacific which beaches were generally dotted with palm trees that leaned toward the ocean; where the sea flowed gently on shore in frothy waves onto a sandy beach; where the thatch-topped huts of the former villagers still stood. Ah, yes, the ideal of a romantic Pacific island . . . except that this was a time of war, and there were no dancing girls or luaus or nights around a blazing bonfire over which was roasting a barbequed pig. In fact, there were no women on the island at all. There were just sailors and more sailors . . . and beer joints.

What Mog Mog had of interest to me wasn't its balmy beaches and swaying palms, but the big tent under which were crates and stacks of beer . . . for a price. Jac Kapp and I made the rounds of all of the beer tents. After all, it was our duty to patronize local businesses, right?

Although Kapp and I, along with the rest of the crew, enjoyed our free time on Mog Mog, it wasn't all a Pacific island vacation. There were still the war games, the practicing to join the fleet at Okinawa. The battle there had been going on since April, and fresh troops and ships were needed in the push to secure the island, the closest yet to mainland Japan. Even though the major fighting had been over on Okinawa since early July, there was still sporadic fighting by Japanese hidden in the rabbit warrens and caves of the island. There was also real and immediate danger to the ships at sea, ships that had to bring in the supplies for the troops on Okinawa.

While anchored at Ulithi, several of our ship's company were transferred to bolster the crew of other ships. On 24 July, Chief Kruse was promoted to Electrician and transferred off the ship. I heard later that, as part of his promotion, he was sent back to the States.

CHAPTER 20
Okinawa

On the morning of 29 July, the crew was called to muster. After roll call, the loud speaker blasted its high warning whistle. Then the Boatswain Mate's voice, "NOW HEAR THIS! NOW HEAR THIS! BATTLE STATIONS! BATTLE STATIONS! PREPARE TO HAUL ANCHOR! THAT IS ALL!"

In the radio room, I went to my station. Garner and Estrada were already there. I said, in passing, "Where do you think?"

Garner looked up. "Okinawa."

"Yeah," I agreed and put on my headset. We had known for a while that we would be joining the fray in Okinawa, but since neither war nor the Navy is an exact science, there was still the slight possibility that we would be headed elsewhere.

The capture of Okinawa had been of the utmost strategic importance as that was the main supply route for the Japanese fleet and mainland, less than 350 miles away. The Japanese had, as early as 1943, prepared their defense of Okinawa called "Absolute National Defense Zone." General Mitsuri Ushijima had the task—the duty—to defend the island against the Americans at all costs. He had at his disposal over 130,000 troops to do that task.

He, like a lot of the Japanese commanders, had a very in-depth battle plan. He decided to concentrate his forces on the southern end of the island, which was easier to defend. This forced the Americans to attack head-on. Aiding Ushijima were the kamikaze planes which he knew would selflessly bomb any ship in the area, with the goal of rendering it useless or sunk.

The stakes in the war had just climbed almost to its peak, for Okinawa was considered an arm of Japan. Although other islands in the Pacific were Japanese territory, Okinawa was different in the eyes of the Japanese. This was Japanese soil. This was the spring-board to the mainland. Whoever held Okinawa held the key to the Empire of Japan. General Ushijima knew that if the Allies took Okinawa they could launch an attack on the mainland of Japan by land or sea. It was imperative he defend the island with everything he had, and then some. He was a cunning commander and, by all accounts, had a deep passion to defend the island under his care. He had pledged to fight to the death, and it was not a pledge he took lightly.

In April of 1945, the Allies had 1,300 ships laying in wait off the coast of Okinawa. Intelligence had reported there was a force of about 60,000 Japanese defenders, when, in fact, there was more than twice that many. The Americans also were not prepared for the "defend until we die" creed of their formidable foe.

On the first day of battle, the Navy landed 60,000 American military personnel at Hagushi Bay, with full confidence that this fighting force could readily take the island. Resistance was expected, but failure was not an option. The Americans felt that they had the superior force, and would soon overrun the Japanese encampments. What they didn't know is that the island was riddled with caves that offered great concealment for the Japanese, plus the amount of Japanese armed forces was far numerous than anyone anticipated. They were dug in and well-prepared for anything the Allies could throw at them.

In May 1945, General Ushijima knew he could not win. The focus of his mission then changed. He must still hold the island at all costs, but now there was a different strategy. For every man he lost, he determined that at least 10 Americans must be killed with him. For every plane shot down, a ship must be taken with it. The Japanese pilots were instructed to aim their planes at a ship when it became evident that they would crash. These tactics proved deadly effective.

By June of 1945, the Americans had taken heavy losses, and more ships and troops were needed to achieve the goal of taking that island, and thus cutting off the main supply route to Japan. As at Iwo Jima, there were also airfields at Okinawa that the Americans needed for their advance on Japan.

Early in July, facing the loss of the island he had fought so hard to defend, General Ushijima committed suicide. It was considered an honorable act.

So it was that the USS *NAPA* joined the convoy steaming toward that embattled island in the Pacific. The ship edged out of the waters off Ulithi with the convoy and headed toward Okinawa. By the time we set sail for Okinawa on 29 July 1945, the Battle of Okinawa was hot into its third month. The land battle was won three weeks earlier, but the fighting at sea continued.

It took eight days to reach Machinato Anchorage, Okinawa. The land battle was all over, except for the mop-up and a few Japanese stragglers. There remained the tremendous task of coaxing the remaining Japanese out of their holes and hiding places, while at the same time reassuring the civilians that they were safe.

The sea battle still raged because of the kamikaze. The crew of the USS *NAPA*, and those of the other ships in the convoy, were ever alert for the deadliest of the weapons in the Japanese's arsenal: the kamikaze, who seemed to swarm above the skies at Okinawa.

Later, it was learned that many of these Japanese pilots were just young boys taught to fly the plane one way to be used as a deadly missile aimed at the ships below. The planes they were put in were hardly more than cardboard or tin boxes, rigged to fly one way to their deaths. To Americans, it seemed cruel; to the Japanese, it was a point of honor.

Almost immediately upon anchorage in Okinawa, the ship's siren sounded. "BATTLE STATIONS! BATTLE STATIONS!" Blasted out of the address system. I grabbed my battle gear and ran into the radio room. The room was already crowded, and I relieved one radioman who dashed out on deck. There were Japanese planes overhead, and at any moment any one of those planes could turn into a suicide bomber, or kamikaze. The big guns on deck started firing. The sound was deafening!

Okinawa was within easy flying distance by the Japanese planes launched from Japan itself. The amount of planes defending the island had made a successful attack on it very difficult, and increased the casualties for the American troops and ships and its allies. Its proximity to the mainland made it imperative to hold the island, for either side. There were four airfields on the island, all of them coveted by the Americans as launch pads for the inevitable attack on the mainland of Japan. They had been fiercely defended by the Japanese, who needed them for their main supply line for the homeland.

At 1620 on 5 August 1945, we started disembarking troops into the landing crafts for transport to Okinawa. For eight hours, the troops went over the side of the USS *NAPA* into the waiting boats. All the while, the air battle waged on around us. The air was filled with the buzzing planes. The Japanese planes zoomed overhead, as thick as bees over a hive. Those which were equipped with weapons fired their guns as they came. For the most part, they were unarmed, except for their bombs and the plane itself. When they were hit by the antiaircraft fire from one of the ships, I watched as the planes nosedived toward a ship. The big guns on the USS *NAPA* kept firing and firing, turning on their swivel to strike at as many of the Japanese planes as possible. The area was so thick with ships that it seemed like they were ducks in a barrel for the airborne missiles that damaged planes became. Those suicide planes, which numbers grew as the war progressed, and the allies came closer and closer to mainland Japan, were already the cause of the sinking of several ships off Okinawa. There was no defense against a pilot who used his plane itself as his weapon, who was willing to die— and had already pledged his life. Even critically injured, a kamikaze pilot would steer his doomed plane into a ship with his last breath. Unlike American fighter pilots, those Japanese pilots did not bail out of the ill-fated planes.

Planes darted about, thick black flak from the antiaircraft guns filling the air around them. When hit, the spiraling plane twirled in the sky like an acrobat pilot at an air show, only this was no show and the pilot, far from entertaining a crowd, had the deadly purpose of destroying the ships below. Those that missed their target slammed with hard, brute force into the ocean. Smoke, fire and swirling waves rose up around downed planes when they hit the sea, sending debris scattering outwards.

Over Okinawa, almost 200 kamikaze attacks were launched against the American fleet and her allies. Of those, nearly 170 were destroyed. Yet, for those that did get through, a monstrous amount of damage was done, especially to the carriers which did not have armored flight decks, such as the British ones.

On 6 August 1945, at 0207, the air raid warning sounded! Every time we had an air raid, we made smoke, or put up a smoke screen, which made it harder for the kamikaze pilots to spot us. There were so many ships out there that if the suicide planes just aimed for the sea, chances were they would smash into an American ship. Although the USS *NAPA* did not get hit, the crew didn't get any sleep either. The following night, the air raid warning sounded again, and once again, I was at battle stations. In the dark, the suicide planes were hard almost impossible to see until it was too late. They zoomed in with no running lights, their light props hardly audible above the flak-flak of the guns and the churn of the ocean. Their sole intent was to crash into a vulnerable spot of a ship. The conning tower, where the radio room was located, was a favorite target because of its height and visibility.

Each time the air raid warning sounded, I could be sure that the kamikazes were overhead, searching for a target. When I was inside the radio shack I didn't see the kamikazes, but I knew they were there. I heard the whine as they flew past, sometimes so low to the ocean that the waves splashed against the underbelly of their planes. The low flying ensured that their planes would hit the hull of their prey, an American ship. I knew they were looking for us, for any American ship. Each time I heard an explosion, it signaled another ship hit, or another plane downed. I always prayed it was the latter, although too often it was the first. Black, billowing smoke rose and bright fire balls from the decks of the ships that fell victim to the kamikazes.

Seaman 1st Class Benjamin C.F. Schlabach, rushing to his battle station, took a shortcut and fell into the open No. 4 hold. He fell four decks straight down, landing on his head. He was pronounced dead less than an hour later. His startled cry when he hit wasn't heard over the noise of battle.

On 8 August 1945, red alert again! There was a lot of **ack-ack** action in the distance, but, thankfully, nothing close to the ship. Billows of black smoke rose in giant puffs as the big guns fired. I stood battle stations, constantly waiting. During the rest of the day into the early hours of the morning, the battle stations blasting signal sounded again and again and again! I could see the flak in the distance; I could hear the whining of the bombs; and I could see the flashing of the explosions all around us. Pillars of smoke rose from decks of ships hit by the suicide bombers.

Finally, in the early hour before dawn, around 0600, all of the cargo was unloaded amid a choppy sea and the uncertainty of attack by the kamikaze. Ship-to-ship communications was at its peak in that time before dawn, all in code that we were sure the Japanese had not broken. There were no lights on deck and no running lights. No smoking a cigarette outside. I was acutely aware that the USS *NAPA* was in the same kind of position as it was off the coast of Iwo Jima when it was rammed. Mentally, I held my breath that another ship of the convoy in the tight quarters of the sea off Okinawa would not make the same devastating mistake.

When all the cargo had been disembarked, it was time for the troops to go ashore. Daylight had come full force by the time they had all gone over the side to the waiting landing crafts, and by 0900 the USS *NAPA* had discharged all of its troops. During all that time, the deck guns fired sporadically as the last of the kamikazes made their final effort to sink or damage American ships. The air was full of puffs of wind borne black flak, so thick the blue sky was obliterated.

Later that day, there was a military funeral for Schlabach, a grim reminder that we had not yet left the war, and of the fact that even though USS *NAPA* and her crew had been spared the wrath of the kamikaze, we were still vulnerable to the dangers of war. After the service, Schlabach was shipped ashore for burial in the government cemetery. Again, we were leaving a crewmate behind.

By the time the battle for Okinawa was over, a heavy toll had been exacted on both sides. 36 American ships were lost to the kamikaze, and 368 more were damaged. Over 750 American planes were destroyed. In the ground battle, more than 7,000 Americans died at Okinawa and another 32,000 wounded. At sea, the terrible tally was even greater: 5,000 men died in the waters off Okinawa, and another 4,600 were wounded.

The Japanese had suffered terrible losses, also. Over 107,000 had lost their lives there, and another 7,400 were taken prisoner. That did not include the civilians, and the toll on them was great. They were subjected to propaganda by the Japanese that the American invaders were barbarians who would slaughter them and eat their children. Also, the Japanese did not look kindly on the natives of Okinawa and, as the land battle progressed, the civilians hid in caves and recesses. The Japanese needed those places for themselves to hide and for sneak attacks against the advancing American troops. The civilians were forced out of their refuges into the path of war. By the time the land action was over nearly a third of the native population had perished from the ravages of war.

The remaining civilians, once they were convinced that the Americans were not the butchers they were warned they were, surrendered. They were in dire straits—covered with filth and lice from their hiding places, and emancipated to the point of starvation. The rice fields, their main stable in their diets, were burned, some deliberately by the retreating Japanese and other rice paddies fell victims to the bombs. The condition of the civilians was a poignant reminder that war is about more than battles between opposing armies.

CHAPTER 21
Philippines

On August 8, the orders to depart Okinawa came while I was off-duty but standing watch out on the deck. There was no such thing as casual time off during active engagements. The USS *NAPA* joined other ships that broke away from the cluster around Okinawa and set to sea again. The next anticipated destination would be mainland Japan.

Later that morning at 1101, we were underway, en route to Saipan, for refueling and to take on supplies. I also expected we would be getting our next orders once we reached Saipan. Although kamikaze attacks were remote on the trek from Okinawa to Saipan, the gunners were nevertheless on alert for stray Japanese planes. It was not time to relax our guard. The sea rolled and the USS *NAPA* lurched with it, but the skies were bright and clear. As we got further and further from Okinawa, I begin to believe that there would be no more attacks from the suicide planes.

Six days later, on 14 August, we dropped anchor in Saipan in the Marianas Island chain. The secure base was where we would take on much needed supplies and fuels. It was also a place where we could, for the moment, relax from the vigilance of war. I also knew that there would be only a brief respite. I was sure we would be in the convoy on its way to the Japanese mainland. The invasion of Japan promised to be far more costly than Iwo Jima or any other battle. None of us looked forward to it, but we all knew that it was inevitable. With the kamikaze as an example, we knew it wouldn't be easy, that the Japanese would fight to every last man, woman and child to save their homeland. We held our collective breaths as we waited for the anticipated order. Iwo Jima was a walk through fire; Japan would be a walk through hellfire itself.

Ideally, the taking of the Pacific islands would have led to the defeat of Japan. But that did not happen. The Japanese were a determined foe, passionately dedicated to continue to fight even when defeat was predictable. There had never been a opposing force with such unstoppable resolve and the unflinching willingness to fight and die in spite of long odds against them.

The following day on 15 August, 1945, at 0945, I was off-duty and standing on the top deck, leaning on the railing, when I expected the confirmation that we would be on our way to Japan to be announced. The piercing whistle of the coxswain mate sounded. "NOW HEAR THIS! NOW HEAR THIS!" I tensed. This was it! We were going to Japan! There was no doubt in my mind, or in anyone's mind, that the invasion of Japan would be costly in American lives, as well as the lives of Japanese civilians. It would be hard fought and become the bloodiest and most costly battle of all. I put out my cigarette; field stripped it without thinking about it, and tucked it in my pocket. I waited, hands unconsciously gripping the railing. The USS *NAPA* and its crew had been lucky so far, but as part of the invasion force, could our luck hold?

The announcement was far from what we expected. "NOW HEAR THIS! NOW HEAR THIS! AS PER ALNAV NUMBER 194, THERE HAS BEEN A SUCCESSFUL CONCLUSION OF THE WAR AGAINST JAPAN. THAT IS ALL!"

At the end of the brief official announcement, there was a horrendous noise from the crew—a thunderous mixture of cheers, tears, and just general ballyhoo! The fire of war was snuffed out! Whatever happened next was going to be a piece of cake!

Just like that! The War was over! I dashed into the radio room, still stunned, unbelieving. The shouts when I entered the small, cramped room reassured me that I had heard right. The war with Japan was over! I squeezed by everyone, high-fiving on the way, feeling an euphoria that is hard to describe. The war was over! No more kamikazes! No more landing on Pacific islands! No more ships sinking around us! The war was over! The jubilation throughout the ship was deafening! The big guns fired, but this time in an announcement of victory! Down on deck, crewmen hooked arms and did a little jig on deck.

I didn't know why the seemingly unyielding foe finally yielded, nor did I particularly care why. All that mattered was that the war with Japan was over! We would not be involved in that final conflict that would easily costs millions of lives on both sides.

What I didn't know then, but would learn in startling, raw detail later, was that while we were off Okinawa, an atom bomb was dropped on Hiroshima, Japan on 6 August 1945. Even with the horrifying result of over 140,000 dead and the city in abject ruin, Japan still did not surrender. Three days later, on 9 August, a second atomic bomb was dropped on Nagasaki, taking with it another 80,000 lives, and almost completely destroying another major Japanese city. Japan finally had enough, and its Emperor Hirochito conceded that it could not win. Their "at all costs" edict had not included the awesome power of the atom bomb. The dropping of the mighty atom bombs had not been an easy or impulsive decision, but well thought out when it became apparent that nothing less than that would convince the Japanese they were fighting for a lost cause. President Truman made the agonizing decision after exploring all possibilities and knowing that, without it, an invasion of Japan itself was imminent. With that invasion, a greater toll would be extracted than with the dropping of the bomb on Hiroshima. The sparing of American lives, along with drawing to an end the war, were pivotal in his making his decision. It was not easy for him, but he knew that it was the lesser of the two world-changing decisions before him: to invade or to unleash the power of the atom bomb. While he had hoped the dropping of the one bomb would be enough of a convincing display of force, he was prepared to drop another if it did not dissuade Japan from continuing the war.

At the time, I didn't care WHY the war was over. I was just thankful that it was! The invasion of Japan had been clinically predicted by the war room analysts as a "high end casualty count," numbering in over a million American lives lost, not to mention a million more Japanese, both military and civilians. No matter why Japan changed its mind about continuing to resist, I was thankful they had. I am certain to this day if the USS *NAPA* had gone on to Japan as part of the invasion force, as it was scheduled to do, I, and most of my

crewmates, would have died there. It was not being morbid; just
the reality of taking the war to the homeland of Japan.

With the surrender of Japan, the mission of the USS *NAPA*
changed. Once again, we became a transport for the Army, this
time to take them to be replacements to Manila, the Philippines. I
watched them come on board with a sense of satisfaction. Instead
of ferrying them into battle with the unqualified certainty that
many would not be returning, the USS *NAPA* was taking them to a
friendly base.

Two days after the announcement of the surrender of the Japanese, on 17 August, we were headed out of the waters off Saipan to
the Philippines. This was further confirmation to those of us who
were still reeling from the announcement that it was true. There
was victory at sea! The war may be officially over, but there was still
much to be done before the USS *NAPA* could pull into home port
and her crew was returned stateside.

Our first port-of-call in the Philippines was Leyte on 21 August
1945, where less than a year before, in October 1944, the Japanese
Vice Admirals Kurita and Nishimura launched a fierce battle to
keep the island. That resulted in the greatest naval battle the world
has ever seen. The two striking forces were overpowered by the
US fleet. The Japanese super-battleships YAMATO and MUSAHI
were there, facing off two Iowa Class American battleships, the
IOWA and the NEW JERSEY. During the battle of Leyte, the scars of
that battle still remained in Leyte harbor. The duel would leave the
MUSAHI dead in the water, leaving Japan without one of its most
lethal battleships.

The Philippines was a crucial location for the Japanese, who
had invaded and captured the island and enslaved its inhabitants
in December 1941. Japan had timed the invasion of the Philippines
with the bombing of Pearl Harbor, and on 8 December 1941, began
its aggressive attack to take over the islands, especially Luzan on
which was its capitol, Manila.

The Philippines was also the sites of the infamous Bataan Death
March in 1942.

The liberation of The Philippines came in March of 1945, by
which time over 71,000 Japanese soldiers, under the command
of General Tomoyuki Yamashita, had died. By the time the islands
were completely secured, the US Army had lost over 3,500 men,

not including those who died at Corregidor and Bataan.

The site of the largest Naval battle in the Pacific still bore its scars that 25 August when we proceeded to Luzan Island, Manila, steaming through historic waters. On our way, we sighted in a distance Mt. Bataan, and the Bataan Death March immediately came to mind. There was Corregidor Island to the port, and El Fraile (Fort Drum) to the starboard. The fierce naval and ground battles for these islands had cost thousands of American and Filipino lives. This was one of the most strategic points for the Japanese, so close to their homeland. This is where General MacArthur made his famous promise ("I shall return!"), and later kept it.

The sight of battle-scarred Manila was testimony to the determination of the Japanese. The once-beautiful city was reduced to rubble. Nothing was spared, from grass huts to marble and brick churches—all was destroyed, tumbled masonry and shattered glass. This was part of the price of war. Looking at the buildings, some which identity were now indistinct among the piles of brick and mortar, I could only imagine the high cost in lives that went along with the horrendous collateral damage. Columns of once magnificent churches lay in crumbled ruin; walls of villas were riddled with shrapnel; deep, once elegant balconies hung precariously over archways that were bent into unfathomable shapes by the fierce attacks on them.

I saw firsthand the war wreckage of a modern city. It was not a pleasant sight. War is not confined to battlefields, nor is it a tidy package restricted just to the soldiers who fight it. Defense plants and associated industrial sites are not the only ones who get the business end of a bomb. War spills over into the civilian population and non-military structures are just as likely to be destroyed. The devastation in Manila was a poignant reminder of how real and terrible war is. Entire residential neighborhoods and business districts had been wiped out. Shells of buildings stood where churches had once been.

In Manila, black market prices discouraged souvenir buying; poisonous liquor discouraged even the lightest of drinking. In seeing the destruction that was once the beautiful and historic city of Manila, I did not feel much like souvenir hunting or even drinking. The results of the Japanese invasion of that place left me heartsick. There was no nice and painless way for war to come to a city. If this much rubble was all that was left of a city like Manila after being in the middle of the conflict, I could only imagine how it was in England after the Blitz by the German planes bombing that city.

The Filipinos were very welcoming to us Americans. They saw us as liberators and heroes. It was here that General MacArthur had lived and been the military advisor, until ordered from the area by the president. President Roosevelt feared that MacArthur would be a tempting target for the Japanese to either kidnap or kill, and he couldn't risk either happening to the General. It was then that MacArthur made his promise of returning in triumphant.

As I walked around the destroyed city, a barefoot little boy, whose pants had holes in the knees, and whose dirty white tee-shirt was sizes too big, put a grubby hand on my sleeve. "Sailor, pom-pom my sister? One dollar! You do, Sailor?"

I cringed in horror at this little boy who was trying to sell services of his sister for one dollar. Yet, who could blame him? Their whole livelihoods were destroyed in Manila. There was no running water, no industry, no hotels, no restaurants. All of that had been taken by the bombs, the mortars, the fighting of the war.

The little boy touched my sleeve again, "Pom-pom my sister, Sailor?" He looked so innocent that I wondered if he knew what he was saying.

I dug in my pocket and dragged out a dollar. "Here," I said. "No pom-pom."

The nameless little boy grabbed the dollar and ran down the scarred cobbled street. I saw him approach another sailor. He was undoubtedly making the same pitch.

I turned and went down a rubble-strewn alleyway to avoid him and others like him, only to run into a boy with a dirty rag slung over his shoulder and a broken shoeshine box held in one hand. "Shoeshine, Sailor?" He said. "Shoeshine, one dollar?"

I looked at the greasy rag he held, then into his eager, hopeful eyes. If that rag touched my shoes, I'd have to toss them overboard and buy new ones in the ship's stores. Yet, how could I refuse him. At least he wasn't selling his sister. I pulled out another dollar and handed it to him. "Not now," I said. "I'll be back later. Okay?"

"Hokay," he said, and pushed the dollar into the bottom of his shoeshine kit.

It was like that all over the city. Little boys and bigger ones hawking whatever they could, especially their sisters to the "soldier men." Young girls barely in their teens, if that, openly soliciting on the street, their young eyes already looking old. They had been through so much and they were struggling to survive anyway they could.

It was here we again lost crewmates, but this time for them to return home. Fifteen of them left for the States, and we wished them well.

The USS *NAPA* was not so lucky as to be ordered stateside. Once the Army troops from Saipan had disembarked from the USS *NAPA*, we moved to berth at Pier No. 1 in Manila Inner Harbor to take onboard troops of the 43rd Division of the Eighth Army.

CHAPTER 22
Japan

On 3 September, the USS *NAPA* took on its cargo and supplies, including military equipment as well as stores (i.e., food, clothing, bandages, medical supplies). The next afternoon, after all the supplies were loaded and stored, members of the 43rd Division of the Eighth Army began coming onboard. These troops, unlike the ones we left at Manila, would be going to unfriendly territory as occupations forces. We were off to Japan!

The big difference between this trip to Japan and the one we feared was that we were not going into battle, but would be there as part of an occupying force. We knew better than to expect the warm welcome we had at Manila. A defeated people do not become friends with their conquerors overnight, if ever. The official war may be over, but the hostilities of the individuals (on both sides) was far from over.

The following afternoon, on 5 September, the fire alarm went off. There is nothing scarier aboard ship, even in port, than a fire on board! I was down at my bunk, getting ready to write a letter home when the clanging of the fire bell blasted. I dropped the tablet on which I was writing, grabbed my helmet and went topside. By the time I got on deck, the "ALL CLEAR" had sounded. I grabbed the arm of a passing sailor whom I did not know and asked, "What happened?"

"Fire in the incinerator room," he said. "Nothing serious."

Relief flooded me for I was already imagining the ship, after surviving Iwo Jima and Okinawa and the kamikaze, going up in flames in a friendly harbor.

I started back down to my bunk to finish my letter to Mom. I caught Jack Kapp on the walkway going down to the lower deck. "Hear what happened?"

"Yeah," he said. "Much ado about nothing."

"Thank God!"

"You got that right. Heard the laundry and stores got a bit smoke damage."

"Not the refrigerator this time?"

Kapp laughed. "No way!"

We both went on down, thankful that the fire had been minor. It was still a pointed reminder that we were not yet out of the woods, and wouldn't be until the USS *NAPA* sat in friendly waters off the coast of the United States. I couldn't help but wonder how long that would be. I wished I could tell my Mom what was happening, but even though the war was over, there was still a great deal that was considered too sensitive for the mail.

On 7 September 1945, we were underway to Yokohama, Japan. It was unreal! Here we were on our way to the harbor of our enemy. The war may be over, but we knew that anything could happen. At 0745 the next day, a most amazing thing happened, which was symbolic that the war in the Pacific was over. All the ships in the convoy turned on their full navigational lights. After so many months of blackouts, the lights looked like an ocean full of ferry boats. The ships' horns blasted, one after the other, greeting one another, free from the need to run silent.

Five days later, on 13 September, I was out on deck when I saw Suno Saki Point, Japan! There is no way to fully describe the feeling when the USS *NAPA* pulled into Japanese waters after all that time, after the walking through the fires of Iwo Jima, surviving the kamikaze of Okinawa, and being amid the destruction of the Philippines.

At 0645, we stopped for the squadron flagship to take on a Japanese pilot to guide us through the port and around the mines that remained in the harbor. I know that the Japanese placed honor above even their lives, but still I was nervous about having a Japanese aboard the flagship telling their Captain how to avoid the mines. I remembered the kamikazes and the stories about the banzai attack. These people considered it an honor to die for their country and take as many of their enemy with them. What would keep the Japanese guide from steering us right into the heart of the minefield?

The mined harbor was one obstacle still remaining to successfully dock in Japan. A few months before, none of us would have taken any bets on our steaming into Yokohama in full daylight with nothing but friendly planes overhead and not a shot being fired. There was something totally surreal about the prospect, but here I was, a part of the convoy heading into the shores of Japan. It didn't seem all that long ago that I had been on the black sands of Iwo Jima, part of the fighting force against these people.

There was a small lighthouse in the bay. It was just a white, round, little, single column sitting on a round base anchored to the sea floor. It was more a beacon than a lighthouse. Someone from aboard the NAPA LCVP painted "NAPA 157" on its side in black paint. We had put our stamp on the homeland of Japan. The simple, and perhaps childish, act relieved some of the tension onboard. We were not here to do battle. The predicted million lives would not be lost on either side. We were here to claim the little lighthouse in Yokohama Bay!

As the USS *NAPA* passed the lighthouse, the railing was lined with the crew and the soldiers we were bringing here to occupy Yokohama. When I saw the dripping painted sign on the wall of the lighthouse, I whistled through my teeth then cheered. Soon my impulsive act was picked up and copied by others leaning against the rail. That little lighthouse was a spot of comic relief for a very tense crew and its company of soldiers.

In the bay were fishing junks, the fishermen aboard and their families, standing in the boats, watching as our ship glided by them. An old fisherman wearing a wide-brimmed straw hat, leaned on his long oar, his eyes boring into mine as the USS *NAPA* went slowly by. He said nothing, did nothing, yet I felt the animosity as real as though he had thrown his oar at me.

The weather was slightly overcast with a haze blurring the shoreline. It was not exactly the most welcoming of weather, but it seemed appropriate for entering hostile waters. Large white squares among the green of the woods showed where coastal defense guns had been surrendered to our advance forces.

At 1046 we passed Fort No. 3: at 1054 we passed Fort No. 2. I don't know what happened to Fort No. 1. By the order of these forts, No. 1 should have been less than 10 minutes beyond Fort No. 2, but it was not there. It was just one of those little things that I noticed to keep my mind off the fact that these waters were mined against invasion, and I wondered again if the Japanese pilot aboard the flagship could be trusted to guide us around the mines. Those forts we passed were the Japanese front fortifications against attack. They were grim reminders of part of what the invading forces would have had to face if the invasion of the Japanese mainland had gone forward. Fully armed, they would have had machine guns pointed at ships entering the harbor, plus antiaircraft guns aimed at our planes. The forts were made of thick concrete that would withstand the first onslaught of bombs or mortar shells. I knew from experience that the Japanese manning those fortifications would have stayed there, fighting fiercely, until the last one drew his last breath.

I went into the radio room as the USS NAPA moved in to Berth A in the North Dock to unload. It was there I learned why the Japanese guide did not betray us and send us into the mines. Emperor Hirochito had called for the war's end. He had issued a statement that the war was over and for all Japanese to cease fighting. The Japanese people respected their Emperor and obeyed him. There was a small pocket of resistance, but they were overcome, and the Japanese people accepted surrender. It would have been dishonorable for the Japanese pilot to have gone against the Emperor and lead the ships into the mines.

I went back out on deck in anticipation of going ashore. There were Japanese civilians on the dock, standing way back, watching. They were quiet. They stood and watched without moving, without talking, without voicing what must have been their distrust and dislike of the conquerors. It was eerie.

That day was spent off-loading supplies. By night fall, the cargo was all on the dock, and guarded by a handful of sailors from all the ships. The crew of the USS NAPA stayed aboard.

The next morning at 0715 on 15 September 1945, the troops we had ferried from the Philippines debarked. They filed down the ramps onto the dock under the watchful eyes of not only the Navy guards but also the Japanese spectators who had returned to the dock earlier. It didn't take long for the troops to go ashore. It was a little strange watching them in their full battlegear, march off the ship onto the land of our enemy. When they had departed and met up with other troops from the other ships, they joined their commander and began their tasks as occupiers of Japan.

It was time for shore leave. I went down to my bunk, donned my best uniform, polished my shoes, and went back topside.

Up on deck, the Lieutenant was standing near the ramp. I saluted him, then turned and saluted the flag. Then I went ashore in Yokohama, Japan. What a feeling! I was ready for a good time, see the sights, and collect a few souvenirs. It wasn't to be. As I walked from the dock into the city, the Japanese people were lined up on both sides of the street. I walked down the main street from the jetty landing and saw all of these Japanese on either side. They didn't look happy to see me. It was like going through a gauntlet at a hazing. Even though they didn't say anything and didn't do anything, I felt their hostility as though it was blazing sticks and pointed stones hurled at me. The city of Yokohama lay in ruins. As a strategic seaport from which the Japanese fleet was launched, it had been mercilessly bombed. Some of its superstructure had escaped but, like with Manila, much of its buildings were in tumbled heaps of brick and mortar. There were no tourist shops, or picture ops, or souvenirs. There was just more evidence of the destruction of war . . . a necessary by-product, but, nevertheless, one that was not pleasant to behold. I sure wasn't happy to be there.

I turned around and went back to the ship and didn't go ashore again. It was hard to separate these people from the enemy whom I had fought back on Iwo Jima. That trip into Yokohama was just one month after the surrender of Japan. It is easy to assume that the Japanese people were not happy with the way the war ended, and surely not happy that their homeland was invaded by those they still considered their enemy. I was part of that group, and it was not hard to imagine how I would be received. I will say, however, that many of our ship's crew did enjoy their stay there. We had strict orders regarding our behavior while onshore. I won't speculate as to how many adhered to it, but there were no hostile incidences committed by any member of the crew of the USS *NAPA*.

The rest of the stay in Japan I spent catching up on my letter reading and writing. I played a few games of cards with some others who felt as I did and decided not to venture out into the streets of Yokohama. It wasn't bad, staying aboard and having free time.

CHAPTER 23
China

On 19 September 1945, we left Yokohama for Guam. I was just as happy to be steaming away from Japan. Once we were in the open sea, we found ourselves in the edge of a tropical storm. The winds blew at 52 knots (65 mph) with gusts up to 70 knots and heavy sea. The rain slashed down at us and the seas swelled up, twisting the ship in an almost impossible curl. The typhoon sent the ship up to a crest on a wave, then the ship would **BANG!** at the bottom as the wave rolled away from, and out from under, the ship. This up and down slamming continued throughout the night. These were shark-infested waters. Any hapless crewman to go overboard was shark food! Hanging onto the railing, leaning into the wind, was the only way to navigate the upper deck.

On 23 September, we pulled into Guam, and docked at Apra Harbor to load cargo for the Sixth Marines Division. We spent seven days at Guam, loading cargo to replenish our own supplies, as well as materiel for the troops, as well as taking on more Marines. Even though there was shore leave permitted, I did not go ashore there as there wasn't really anything to see, except more destruction and more grave sites. I had seen enough of both. The cost of winning the war was high; but the cost of losing would have been indescribably higher. There was never any choice but to fight to win.

Just a week later, on 30 September the USS *NAPA* was underway again, this time heading once again to Saipan. By 1 October we were in the waters just outside the harbor of Saipan. It dawned on me while the ship slowed outside the harbor that I had been aboard the USS *NAPA* exactly one year to the day. The paravanes were streamed as mines were still a possibility outside the harbor. The paravanes were torpedo-shaped devices with sharp fins at the front, that, when towed by the ship, cut moorings of submerged mines, sending to the bottom of the ocean instead of into the hull of a ship. Even though Saipan was a friendly port, long taken by the Americans, the threat of undetected mines in the surrounding area was very real. There were some tense moments while the paravanes were towed. When cleared of the outer harbor, the USS *NAPA* docked at Berth 44. We only spent one day there, unloading Marines. No one other than the Marines went ashore.

In China, across the Yellow Sea from what is now North Korea and just south of Manchuria and the Gulf of Po Hai Gulf (Chihili) lies the port of Chefoo, China (now called Yantai). It was there the USS *NAPA* headed after Saipan. After a brief refueling stop on 10 October, the USS *NAPA* headed south to Tsingtao (now Qingdao).

On 11 October 1945, we anchored in Tsingtao Harbor in China, up near the Russian border. This town was international, having the past influences of German, Russian and Chinese. It was an ultramodern city for the times and had been spared destruction due to its proximity to Russia. This was a friendly harbor, since the Chinese had just been freed from the slavery yoke of their previous Japanese masters. The Japanese had occupied and brutalized China since 1937, and it was only with the dropping of the atom bombs on Hiroshima and Nagasaki, and the subsequent surrender of the Japanese to the Americans, that China became free of its Japanese slave-masters. After the horrors through which the Chinese went during the occupation by the Japanese, they were overwhelmingly grateful to their American liberators. They knew that even though the Americans hadn't actually set foot on the soil of China to fight off their captors, it was because the Americans had defeated Japan that they were free. The Japanese had withdrawn when word came the war was over. While there were a few pockets of resistance, Japanese who refused to believe the war had ended, all of the occupying forces had left Tsingtao.

A flotilla of sampans came out to greet us, the families standing and waving as the ships made their way to the dock. (Sampans are small, wooden Chinese fishing boats on which the owners lived, ate, slept, and conducted their business.) The fishermen and their families were, of necessity, excellent swimmers since they spent their lives on the water. As we came closer to the harbor, one of the sailors leaning on the rail next to me said, "Watch this!" He gave a shrill whistle, and some of the people in the sampans looked up. He drew his arm back and tossed a coin into the water.

Immediately, several people from different sampans dived into the water. After a moment, a triumphant diver came up holding the coin high in one hand. He stayed in the water, waiting.

I dug in my pocket and came up with a quarter and heaved it into the water. The same diver, and others, went quickly under the water, and one of them came up holding the quarter in his hand. I did it again, only this time with a nickel, with the same results.

Soon, most of the sailors lining the railing of the ship were tossing coins in the water, as little as a penny and as high as a quarter. Every time a coin would hit the water, under would go one of the people from the sampans. One sailor heaved his shoes in the water, and, sure enough, a diver followed it underwater and came up with both of them. It didn't matter what was tossed to them, they expertly dived for it and were always triumphant. At the time, none of us considered it disrespectful, and the sampan divers sure didn't mind getting the extra coins . . . or a free pair of shoes!

Shore leave time! I eagerly went ashore with sixteen other sailors. Tsingtao was a place where Americans were not only welcomed, but also encouraged to stay.

The Chinese were pleased to see us, and welcomed us with opened arms. Tsingtao became known as Utopia, the land where we could get 15,000 Chinese dollars for one American dollar. Every Chinese an American met smiled in welcome. All the streets were draped in paper flags, and the Chinese citizens applauded when American military men went by. There was an unsubdued gaiety that captivated everyone who went ashore. It was a very pleasant change from the other places where I had shore leave. The city was intact and the natives were friendly.

Being in Tsingtao was a completely unique experience to any other place on the globe. The international community had left its mark, not only in the architecture, but also in the foods and wine, and especially vodka, which flowed like a river free of its dam. All the American whiskeys were sold there, along with local fermented rice wine and Russian vodka. It was a place where a sailor on leave could happily have no inhibitions. There weren't any reports of muggings in Tsingtao. The Chinese code of honor would have forbidden it.

There were fine hotels in Tsingtao, fully staffed by grateful Chinese. They were luxury places where the wealthy would have found it most pleasant to stay, or to dine at one of their fine restaurants. I wound up at one of their main hotels to have dinner. Along with dinner, it would have been insulting to my hosts not to try their wonderful, Russian-inspired, vodka.

The main mode of transportation through the winding, narrow streets of Tsingtao was the rickshaw. It was said that the name "rickshaw" came from a Japanese word meaning human-strength-vehicle (jinrikisha). The name was adopted throughout the world where there are human-pulled rigs, although the shape and capacity may be different.

The rickshaw is a two-wheeled cart, with two long handles, which seats one or two people, and is pulled by a runner. The high wheels are spoked, hard-cast, like wagon wheels. The rickshaw runner is the earliest form of a taxi driver. I had gone from the dock where the USS *NAPA* was to the hotel by the rickshaw. I, along with the others from the USS *NAPA*, asked my driver to wait while I dined. He gladly did so. The rickshaws were lined up outside the hotel to take us back to the ship.

When I finished my fine dinner, along with my ultra-fine vodka, I was feeling no pain. My friends were in same vodka-happy induced state of mind. I walked with much determined dignity out of the restaurant, and there was my faithful rickshaw driver, leaning against his rig, waiting for me. The handles of the rickshaw rested on the pavement and he quickly got between them and lifted them, one hand on each handle. I got aboard, and my buddies got settled in theirs. My rickshaw driver, and the others, started up the cobblestone street to the landing so we could go back to the ship.

My buddies started shouting, "Giddup!" The rickshaw drivers either didn't understand or were used to such comments. They didn't seem insulted at all.

We got up to the crest of one of the cobblestone streets, and I was leading the pack. My rickshaw runner was ahead of the others.

At the top of the hill, I hollered, "STOP!" and my rickshaw driver halted, still holding the poles in his hand. He waited for instructions.

"What's up, Geurin?" One of my buddies asked from his seat in his rickshaw.

"Got an idea," I said, and it was an absolutely brilliant idea.

The other rickshaw drivers "idled" at the curb . . . they stood there, poles still held in their hands, feet firm on the street, waiting for directions.

"Hey, buddy," I said to my driver as I got off the seat of the rickshaw, "Get in." I motioned to the seat I had just vacated.

He shook his head and said, "No, Mister. I pull you."

"Come on," I said, tugging on his arm. "It'll be fun." He still looked doubtful. "Hey, I'll still pay you." I could tell he was not convinced, but he reluctantly got into the seat. So, over the driver's objections, I put him in the rickshaw and got between the long poles and started running downhill.

The other rickshaw drivers started up, and my buddies shouted, "Way to go, Geurin!" Going downhill was not foreign to those drivers, and they kept up a good pace.

I didn't realize how balanced the rickshaws were and how fast they could go. I met another rickshaw coming up the hill coming directly at me! I had gathered up speed and couldn't stop or swerve because the other drivers were spread across the road. I hit the oncoming driver head-on! The rickshaw driver fell to the ground, letting go his rickshaw, which overturned with the momentum, and knocked a lieutenant out of his rickshaw. Naturally, I kept going, and I remember hearing the lieutenant's yelling, "Get that sailor's name! Get that sailor's name!"

When I was finally able to get the rickshaw stopped near the bottom of the hill, the Chinese driver got out and forcibly put me back into the seat of the rickshaw. I didn't resist, and we took off again. This time, the driver was running between the poles.

We got down almost to the jetty landing. I jumped out and hastily paid the driver, cramming a bunch of American dollars in his outstretched hand. There were sampans bumping against the dock and tied together for over a hundred yards out. This was a little floating city. I quickly glanced behind me, expecting to see the lieutenant or the rickshaw driver whom I had toppled right behind

me. Nothing. Yet. But I wasn't in the clear. Either of them could still
come dashing round the corner at any moment. Andrenalin pumping, I checked out my options. I was cornered! I couldn't go back
the way I came. The lieutenant and-or the rickshaw driver would
more likely be rushing from there. The ocean was in front of me.
I took a breath and leaped from the jetty onto the first sampam.
The force of my jump sent the sampan to waving, and the owner
popped up from where he had been laying down and hollered at
me, something Chinese that I could only imagine the meaning. I
waved at him and leaped to the sampan next to his, barely missing
a woman preparing her family's meal. She looked up, startled, just
as I jumped to another sampan and another, all of them occupied,
and all the occupants expressed their surprise at a sailor taking jar-
ring leaps across what was to them their homes. It would be like
dashing through the living rooms of one house after the other back
home. The only excuse I had was that I was attempting to escape
from the probing eyes of the angry lieutenant and the toppled
rickshaw driver. I must have run over at least five or six sampans
before I finally jumped back upon the landing to dash down to the
end and get on the boat to take me back to the ship.

Finally, I went back to the ship and went down to my bunk and
fell in an exhausted sleep. I had long since overcome any vodka
buzz! I had successfully evaded detection! My heart had been
wildly pumping, but I knew my buddies back there in the other
rickshaws would never divulge the name of the flying rickshaw
driver who spilled the lieutenant. If my luck held, either the lieutenant wouldn't be from our ship or, if he was, he didn't get a good
look at me.

The next afternoon, on 16 October, 1945, the USS *NAPA* received
additional food stores that we hadn't had in a long time—fresh
eggs! Lee Shun Company of Tsingtao delivered 4,800 dozen eggs
for the men above the USS *NAPA*! I watched as the loading boom
was lowered with the large net to bring them aboard. Lee Shun's
delivery person manhandled the crates and piled the first ones in
the net that was spread on the dock and signaled the ship to "Haul
away!" The first batch was brought aboard, and the crew from
Stores eagerly off-loaded them. The net was lowered again, and the
action repeated. The second batch off-loaded and away went the
net to the dock once more. By this time, the loading of the fragile

eggs was becoming routine. With routine comes carelessness. The net loaded with the third batch of eggs begin its journey up the side of the ship. Then . . . WHAP! The net swayed and bumped and hit the ship with a loud CRACK! I leaned over the side of the railing just in time to see the net banging again against the side of the ship. There, going down the ship in a drippy, slow mess was the result of several dozen eggs smashed. The whites and yellows of the eggs oozed through the net and splattered against the side of the ship, making the USS *NAPA* look like the butt of a Halloween joke.

CHAPTER 24
Formosa

On 17 October, we left Tsingtao, China, for Manila again. It took six days to sail to the Philippines, and on 23 October, we pulled into the harbor. Some of the crew were sent ashore for discharge. I stood there on the deck watching them, wishing I was one of them.

I went ashore only once during the seven days we were in Manila. All there was were the bombed out buildings I had seen before, and the war-weary people. There was really nowhere to go in Manila. It hadn't changed since the last time I was there, and I had no stomach for meeting, again, the young boys selling their sisters for a dollar. We were warned again not to drink the water or the booze, and not to touch the women. It was really sad, the way that young kids would run up to the sailors and offer their sisters, their mothers, their aunts, and themselves for a price. It was a by-product of war, this desperation to sell something, anything, even one's soul, to get money for the black market food, water, and the necessities of life. There was no longer a financial base in Manila. Everything had been destroyed during combat. What was left was desperation and the seedy black market operators. This selling of the young girls was the result. War does not leave a besieged city intact. It strips everything and everyone of even the basic of human dignities. The difference between Manila and Tsingtao was startling.

On board ship, I was standing watch at the radio shack and

heard moaning outside on the Captain's deck. Curious, I went out and discovered that the sound came from Seaman 1st Class Jack Stanfield, who was also a radio operator. This was the same man who had once challenged me to a grudge boxing match. He was lying there, moaning, obviously in great pain. I picked him up and carried him, fireman's style, down to sickbay. The medics there told me that if I hadn't gotten Stanfield help when I did, he would have died from something he had consumed in Manila.

By 30 October 1945, we were underway again, en route to Haiphong, French Indo-China (which is now known as North Viet Nam). On 2 November, we reached the Tonkin Gulf off Doson Peninsula in French Indo-China. (Twenty years later, on 27 July 1994, the Vietnam War would begin in that very spot.) The Beach Party, of which I was still a part, was sent ashore to coordinate the loading of General Chaing Kai-Shek's National Chinese troops. For the first two days, the other crew members had to stay aboard the ship to see to the settling of the Chinese.

The General's troops were considered allies. During the war, Japan overran parts of China and trampled the Chinese in their wake. The main cities of China were commandeered by the Japanese.

The Chinese were also fighting each other at that time. There were the Communists led by Mao Tse-Sung against the Nationalists led by General Chaing Kai-Shek. At one point, it became evident to both Chinese factions that they had to band together to fight their common enemy, the Japanese. They formed an uneasy and temporary truce.

Mao Tse-Sung had risen from peasant class to become one of the most powerful Communist leaders. He was in direct opposition of the then-leader of China, Chiang Kai-shek, who systematically purged Communists from any office in the government. In the 1930s he led campaigns against the Communists, and Mao was forced to flee to the interior. Both men had devout followers at the time that Japan invaded the Chinese mainland. At that time, the two factions temporarily united against the common foe.

As soon as the Japanese surrendered, the Chinese's shaky truce with each other fell apart. The Chinese Communists went after the Nationalist troops. The Communists not only outnumbered the Nationalists, but they were also better armed. Chiang Kai-Shek's troops were left in deadly peril. It was this danger that prompted the United States Navy to offer sanctuary to the Nationalists Chinese. They were quickly being surrounded by the Communists and had to be gotten out to keep from being totally annihilated.

The Mao Tse-Sung and his Communist party formed an alliance with Russia against the United States, and that was all the more reason for the United States to ensure the safety of Kai-Shek and his troops.

It took two weeks, but 3,000 of the Nationalist Chinese troops were finally aboard the USS *NAPA*. The other ships had also taken aboard massive amounts of the Chinese troops. We ended up with elements of the 62nd Chinese Army on the ship. All the time I was with the Beach Party to assist in the boarding of the troops, I slept off ship. I had the seabag and tossed it on the ground near where the Chinese troops were gathered. Garner, Estrada and I put together our radio and established shore-to-ship communications.

Lieutenant Skubitz, our commander, had a Jeep on shore while the rest of us were walking. He had parked said Jeep under one of the few trees in that area and took off for a latrine call. While he was occupied, a certain Chief came along and decided the Jeep shouldn't be just sitting there, idling, doing nothing. He nodded toward us and said, "Hey! Wanna come 'long?"

I joined the others, and as many of us as could squeeze into that Jeep piled aboard, and the Chief took off! There was nowhere in particular to go, but it was a break from the tedium. He took the Jeep down to the beach, where he spun out in circles. To calls of "Way to go, Chief!" and whistles, he skidded across the sand, bouncing a bunch of us off as he went merrily along. He pulled to a stop, spinning up sand in his wake, and offered the wheel to anyone who wanted it. A few of us took turns, seeing how tight of circles we could make, the rest hooting and hollering and daring.

All of a sudden during this free-for-all, the guys standing around watching became awfully quiet . . . and also started to disperse rather quickly. There was an angry shout, and I started scampering away with the others. "HEY!" the Lieutenant shouted. "What the hell's going on here!"

Somehow in the confusion that followed, the Chief had slipped away. The rest of us? We very docilely followed the Lieutenant back up to where the Chinese Nationalists were gathering to go on board the USS *NAPA*. When we got there, the Lieutenant said, very tightly, "All of you are on report!"

Ouch! That meant no liberty in the next port of call! Turned out he suspended all privileges for all of us for three days. That was all right . . . we were busy getting the Chinese off the beach and onto the ship.

For the most part, the 3,000 Chinese were just kids, younger than I, and I was only 20 at the time. Some of them looked as though they couldn't be more than twelve or fourteen, and there they were, fighting for their country. They were now forced to abandon their homeland, leave their belongings behind, and go, instead, to an island off the coast of China. I couldn't imagine being forced out of my country, especially after having fought for it. It was a political ploy by the Communists, forcing the Nationalists out of the country onto the small island of Formosa. The Communists hoped to break their will, but they were not successful. The Chinese Nationalists held onto their beliefs, even with their backs to the wall.

The Chinese swarmed all over the ship, even though, technically, they were supposed to be confined to one area. There was just no way to enforce that. I discovered early on that if I flipped a cigarette to the deck, several of the Chinese would rush forward to toss it overboard. Their ships had been made of wood, and they were deathly afraid of fire at sea. It soon became a game to us, although it was serious for them. In looking back, I hope they realized that we saw it only as a harmless prank. We did not know what they had been through, which was as terrible as the taking of Iwo Jima, if not worse. We meant no insult; sometimes words and actions are difficult across cultures.

Once all the Chinese troops were aboard, our ship pulled out of the harbor and headed toward Takao, Formosa (which is now known as Taiwan), which is south of the Chinese mainland and north of the Philippines in the South China Sea, and safety for the Chinese. On 18 November 1945, we anchored off shore five miles off north Tako. Shortly before noon, we started unloading the Chinese Nationalists. It took a little over four hours before they were all safely on Formosa. The island country had been taken over by the Japanese during the war, but at the time we ferried the Chinese Nationalists there, only the American military was there. The Japanese had been run from the island during a horrific battle there, and the Americans had taken over.

The Nationalists established themselves on Formosa. Much later, Communist China would declare Taiwan as a providence of China. The Nationalists would continue to resist.

CHAPTER 25
Manila

19 November 1945, we were again going to Manila. The only bright spot was the scuttlebutt that we would finally be heading stateside. It took two days, and by 21 November, we were again anchored in Manila Bay. Although Manila no longer held any charm, and I had been heart and soul sickened on previous occasions by both the destruction and the depravity, going ashore was still preferable to staying on ship. There was no drinking or carousing on Manila—the same prohibitions were still in force about poisoned whiskey and personal danger.

Instead of going on into town as I did before, I chose to stay close to the beach. Several of us did. We got up an impromptu baseball game. The Yankees didn't have anything to fear from our competition, but we did have fun! We sort of made up new rules as we went. It was, by far, a lot more enjoyable than going once again into the destroyed town. I will say that my hit in the third inning should have been a home run. Can I help if the ball got hung up in a palm tree and was ruled unplayable?

There was a USO show that came aboard the ship. I didn't care that the cast were not famous in the category of a Marilyn Monroe or a Bob Hope. It was still a merry show that entertained us, and that was, after all, the point.

A lot of times, we would put on our own shows. There was a jazz band aboard the ship and, naturally, we had our share of comedians.

Some of the sailors even dressed somewhat like women to do their part in our impromptu performances. It was a blast, and one way not only to break the monotony, but also to slowly return to normalcy after the look into the face of hell on the islands.

When I awoke on Thanksgiving Day, 22 November 1945, I had a lot for which to be thankful. The war with Japan had drawn to a successful close. The war in Europe was long over. I had seen death and destruction in many parts of the Pacific, especially on Iwo Jima, and survived to tell about it. I had set foot on Japanese soil, not as an invader, but as part of an occupying force. I had seen burned out buildings and decaying corpses, and heard the last gasp of breath from a Marine on a stretcher, but had seen many more saved. I saw small planes turned into blazing missiles and spiral down to smash into a ship, and saw the miracle of a carrier survive the worst of it. I had looked into the hollowed eyes of youngsters who had everything ripped from them and still had the courage to go on and rebuild their lives. I watched an entire people uprooted from their homeland and deposited on a miniscule island to start all over. I had seen it all, and yet here I was, alive and physically unscathed on Thanksgiving Day out in the Pacific, and so I was thankful.

Later that day, my reasons for being thankful increased tenfold! The shrill whistle the proceeded an announcement blasted throughout the ship. "NOW HEAR THIS! NOW HEAR THIS! THE CAPTAIN IS SPEAKING! NOW HEAR THIS!"

The last time the Captain made an announcement it was the end of the war with Japan, so my heart soared with the thought that this, too, had to be good news. Could we be that lucky twice?

"This is your Captain speaking." He began and the noise aboard the USS *NAPA* lowered until there was an anticipatory silence. "In Manila today, I turned in the Oriental rug and exchanged it for a magic carpet. Stateside! That is all!"

The USS *NAPA* would fulfill its duty once again as a troop transport on 27 November. Only this time, in addition to the 2,000 regular Army troops taken onboard there in Manila, there was a very special contingent of the Army that I, along with every guy on the ship, eagerly awaited. For the first time, the USS *NAPA* would be carrying women. Three WAC (Women Army Corps) officers, seven Army nurses, and 79 enlisted WACs boarded at Manila. I was one of the sailors leaning over the railing watching as these special Army troops came on board. In spite of our well-earned reputation for being restrained, there were a few cat whistles that slipped out. As the women marched smartly onto the deck, in spite of their Army uniforms, there was no doubt of their femininity. Women! Aboard the USS *NAPA*! Now there was a situation any swabbie could appreciate!

Shortly after the WACs were mustered by their commander, our Lieutenant called us to order. There he laid down the rules. "The WACs are strictly off-limits," he said, without preamble. "You are not to go near them. You are not to speak to them. You are not to bother them in any way. When you need to pass them, you will do so without making eye contact with them. That is all."

When the Lieutenant finished speaking, leaving no room for discussion, I walked back to the railing and looked over. So, okay, I couldn't speak to them, but the Lieutenant didn't say anything about looking! I wasn't the only one. None of us had blinders. These were the first American women we had seen in a long, long while, and for a bunch of sailors out at sea, they were a soothing balm to the eyes and to the spirit. They sure looked pretty good to me! Even though the orders were not to look directly at the women, what man aboard ship could obey that?

CHAPTER 26
San Francisco

On 28 November, the USS *NAPA* set off for home! California, here we come! Our orders were to go to San Francisco, where the Army troops, including the WACs would disembark. Visions of riding cable cars, and dancing at the San Francisco Ballroom, and sampling American beer in every bar near the Bay washed over me. San Francisco. Now there's a sailor's ideal leave town.

In the balmy Pacific weather, the women would regularly sunbathe on deck, exposing a lot of long legs and bare backs in their swimming suits. While going anywhere near them was strictly off-limits, and we were supposed to act as though they were not there, to not look directly at them, only a blind sailor could do that! It was an impossible order to obey! I had one advantage in that the radio room was higher than the top deck, so I could admire the view without being too obvious about it. Of course, it didn't take long for others to figure out that the radiomen had a great vantage point, and come and enjoy the scenery, too! There was also a port hole that was strategically placed for our viewing pleasure. It got very crowded in there!

The lookouts in the Crow's Nest volunteered for double duty. Their binoculars weren't always pointed out to sea. In fact, they were rarely pointed out to sea.

The officers had given up their quarters to the WACs. Outside the officers quarters was a narrow passageway. Whenever the

WACs were inside, that passageway became very, very busy. You see, there were portholes in the officers quarters that had shades for privacy. Those shades did not pull all the way down. There was a gap at the bottom that was just enough to make sure the occupants inside were . . . safe, shall we say?

Every morning when it was gauged the WACs were getting ready for the day, the passageway outside their quarters became very crowded with a line of sailors who took it upon themselves to make sure the women inside were "safe." Now and then . . . in fact, too frequently . . . I'd have to kick the guy ahead of me in his shins to get him to move on. After all, it was my duty as a radioman to communicate everything that happened aboard, and making sure the women inside were safe was right there on top of my list of priorities.

On 7 December that year, Mother Nature took care of what the Lieutenant could not manage. Rough seas and cloudy skies and waves washing over the deck of the ship took care of our all-too-brief illicit enjoyment. No more WACs on deck in their bathing suits or, for that matter, anywhere visible at all!

15 December 1945, the USS *NAPA* sailed into San Francisco Bay. The cheers on board were almost as loud as when we had received the announcement of the Japanese surrender. I was one of the first ones to line up on the railing as the USS *NAPA* entered the waters of the San Francisco Bay. The entire ship's company lined up on deck to see the Golden Gate Bridge in the distance. As the bridge seem to grow in size the closer we came to it, the mood on deck subtly changed as though until then we had all been holding our collective breaths, not believing this was happening. When the USS *NAPA* sailed under the Golden Gate Bridge, I looked up and saw the underside of that great structure and it was then it struck me: we were Stateside. Maybe for good??

Finally at port in the friendly city of San Francisco, we were called to muster. I snapped to in my dress uniform, and when roll call was over, the Lieutenant said, "Liberty for off-duty personnel. Starboard. That is all." As was the routine, those with port passes would go the following day.

It so happened I had a starboard pass. But then, I also had a port pass. The fact was, I had a permanent, two-sided pass, courtesy of a certain entrepreneur over a year past. Since all I had to do was flash the pass to get liberty, and not hand it over to anyone, it had served me well. So far. In addition to that universal pass, I also still had my identification that said I was 21, even though I was still only 20. I seriously doubt if many bars stateside would refuse to serve a guy in uniform, but it didn't hurt to have ID that said I was legally of age to drink.

I talked it over with my friend Jack Kapp, and we decided that San Francisco could wait. He and I had a nine-day liberty, and we decided my Mom's home cooking beat anything San Francisco had to offer. Another friend of ours, Beardlesy from a sister ship, met up with us before we took off. He was just as eager as we were to get a taste of home cooking for a while. It was only 285 miles south to Bakersfield, and a sailor's thumb was still his best mode of transportation. There was no shortage of motorists willing to give a man in uniform a ride anywhere! So it was that Kapp and I hitchhiked to Bakersfield. It was joyous, visiting with Mom and Dad. I was in full dress uniform. Back in those years, we didn't go outside without being in full uniform. Besides, I was proud of wearing it.

When the last of the rides let us off in front of my parents' house, I gave a yell, Mom, I'm home!"

Mom came to the door, wiping her hands on her apron. She saw me and cried, "Arvy!" I dropped my seabag and ran up to her and gave her a big hug. It was the best hug I'd had since I'd joined the service! Mom then spontaneously hugged both Kapp and Beardlesy even before I introduced them.

We all three went inside, and Mom had us all sit around the table, and gave us homemade cookies and big, cold glasses of milk. We then stored our knapsacks and I told Mom I was going to show the guys around.

The three of us went to my old high school in Bakersfield and visited my teachers from my Junior year. I chatted for a while with the swimming coach, Mr. Shoop, and learned about the others from my school—both before and after me—who had also gone to war. I found out that the captain of the football team, LeRoy Seeger, had been killed overseas. There were five others of the young men who had attended Bakersfield High School who had been killed in the war. Their names and pictures were listed in the hallway, displayed prominently and with pride and honor. I knew most of them, and it felt sort of strange to know that these friends of mine were never coming home.

Mom's contribution to the war effort was to provide a little home life to visiting servicemen. Mom would go down to the local USO and bring home some of the guys for a little home cooking and relaxation. While we were there, Kapp, Beardsley and I met a Marine who Mom had brought home from the USO. The Marine, who was from New York, made it a practice to stay with Mom and Dad whenever he was on leave. He was just like another son for them. He never got overseas. He was stationed in Barstow, California, and went to Bakersfield as often as he could. I liked him. He was quiet, mowed the lawn, helped around the house, and stayed at home when he was in Bakersfield. I liked the idea that there was someone there for Mom and Dad while I was gone.

The first night of our leave, we decided to go into the USO for the dance. The three of us were in our dress blues. Dad lent me his car, and off we went. It was already crowded when we arrived. There were Marines, Army, and Sailors there. Of course, there were also the pretty girls.

The three of us went to the hospitality table and filled paper plates with doughnuts, and grabbed paper cups and looked around the room. There was a four-piece band playing a slow song, girls sitting on metal folding chairs, chatting to each other or watching the crowd, and men in the uniforms of all the services milling about, stopping here and there to ask a girl for a dance.

Beardsley said, "I'm going to go over and ask that brunette to dance."

We had been watching the girls next to her, and had seen that a Marine had just brought her back to her chair. He was across the room getting the drinks when Beardsley casually went over and asked her to dance.

She got up with him and away they went on the dance floor.

Kapp and I were just deciding which two other girls would be the lucky ones to dance with me when I saw the Marine approach Beardsley. All USO dances were what were known as "tag dances." That is, anyone could cut in simply by tapping the man on the shoulder, and whoever was tapped, or "tagged," was expected to step aside. The Marine tagged Beardsley on the shoulder, and Beardsley stepped aside, but not far. As soon as the two were in full swing, Beardsley went over and tagged the Marine. Kapp and

I could tell from where we were standing that the Marine wasn't particularly liking being tagged, but he did step aside.

I nudged Kapp. "Watch," I said, "I'll bet the Marine cuts in."

Kapp laughed. "No doubt."

Sure enough, the Marine tagged Beardsley, who stepped aside.

This back-and-forth went on for the reminder of the dance set. When the band stopped for a break, it was Beardsley who was dancing with the girl. He led her back to her chair, and came to join us. He was sweating slightly. He scooped up a cold drink in a paper cup and took a big swallow.

"Having fun?" I asked.

"You bet," he said and walked back across the floor to the brunette. This time, Kapp and I followed him.

Sure enough, the Marine came over. The girl, whose name I had yet to hear, excused herself and said she'd be right back. The Marine said, "Feller, I'd just as soon you leave my girl alone."

"Your girl?" Beardsley scoffed.

The Marine nodded.

"When'd you meet her?"

"Tonight," the Marine said, "but I like her, and I'd just as soon you leave her be."

"Tell you what," Beardsley said, "I like her, too, and I just met her tonight."

"Take my advice and leave her alone," the Marine warned.

"And if I don't?"

"Then," he said, sounding serious, "it'll be you and me in the parking lot."

"I don't think the parking lot's such a good idea. Tell you what, how about we take Geurin's car here and drive out five miles and duke it out. The one who wins, rides back; the one who loses, walks."

The Marine thought it over.

"Before you answer," Beardsley went on, "I think I should warn you that even though I'm shorter than you, I was Golden Gloves Champion four years running."

The Marine looked at Beardsley and I could tell he was gauging him. About then the brunette came in sight, coming from the direction of the ladies' room.

Beardsley did a sharp left jab in the air, a little pivot dancing on the balls of his feet, shadow boxing. "How about it?"

The Marine said, "Another time." He walked away and all of us sighed big in relief.

Kapp clapped Beardsley on the back. "You old son-of-a-gun, you!" he laughed. "You've never boxed in your life!"

Beardsley laughed. "Nope, but I can dance!"

The rest of the night, the three of us managed to dance with several of the girls. I don't know what happened between Beardsley and his brunette, but I do know he left her there. He didn't make the same mistake I had with Helen in Seattle.

The rest of our liberty in Bakersfield, the three of us . . . sometimes four with the Marine Mom had "adopted" . . . made ourselves useful around the house. It felt good doing chores for Mom and Dad, chores I used to gripe about to Elton about doing when I was growing up. We took out the trash, painted the porch, mowed the lawn, raked the leaves, and in general made ourselves useful. In turn, Mom made the most wonderful pancakes, smothered with maple syrup and dripping with melting butter, and homemade chocolate cake with thick, gooey icing, and deep hearty stew with thick gravy. I don't think any of us missed roaming around San Francisco with the other guys from the ship. The San Francisco beers couldn't taste as good as hot chocolate at Mom's.

Those nine days flew by fast. Much too fast. Mom and Dad drove Kapp, Beardsley, and me back to San Francisco at the end of our leave. There was a tradition in our family that when any member parted, he or she would kiss Mom and Dad goodbye. I wanted to get something straight and settled before we got to the dock. On the way to San Francisco that night, I was riding in the backseat right behind Dad. Beardsley and Kapp were sitting beside me, and Mom was up front. I tapped Dad on the shoulder and said, "Dad, I think I'm getting a bit old for this kissing goodbye bit, don't you think?"

Dad didn't say anything. He just nodded, and paid attention to the road. There was Toulie fog throughout . . . that low-lying fog that makes it almost impossible to see the road. That wasn't really the time to talk about goodbyes, but with my buddies sitting there, I had decided that I didn't want to look like a sentimental little boy in front of them instead of the man I was when we got to San Francisco.

When we reached San Francisco it was foggy, and I could hear the crying of the steam whistles out in the Bay. It was an eerie and mournful sound. It was night, and I could see the ship from the jetty landing. Dad stopped the car, and all five of us piled out. Beardsley and Kapp gave Mom a quick hug and thanked her for her hospitality. I kissed her on her cheek and hugged her tightly. Beardsley shook hands with Dad and stepped back, then Kapp did the same. Dad held his hand out to me. Just then, the fog horn sounded again, long, wailing like a banshee. I grabbed Dad's hand and pulled him to me and kissed him goodbye. He nodded, patted my cheek and he and Mom got back in the car. I watched as he backed up, turned around, and drove away. I stayed there until the red of his taillights disappeared in the fog.

I never felt so lonely in all my life as when I stood there waiting to go back on board. It had been a good leave, but now I was going back to wherever the USS *NAPA* would be sent next.

Beardsley shook hands with us and hurried down the dock to where he could catch a boat back to his ship. Kapp and I got into the LCVP (Landing Craft—Personnel) and headed out to the ship. Once onboard, I walked to the bow and just stood there looking out into the fog, wishing I wasn't going back out. I know Kapp felt the same, as did all the sailors who returned from their leaves in San Francisco or anywhere else.

CHAPTER 27
Shanghai

On 4 January 1946, orders were to sail to Shanghai, China, and the ship eased out into San Francisco Bay to the Pacific. I was going halfway around the world again. On 5 January, we changed course and returned to San Francisco under orders. My hopes soared that we were not going overseas again. I had seen enough of over there. Hopes were dashed the next day when we changed course once again back to Shanghai. This time the orders stood. This time we were not in a convoy. We would sail alone to the shores of China. Sailing alone, without the escort ships or being in the company of other troop transports, felt like going naked in Times Square. I felt we were very exposed. It was taking some getting used to, this not being in battle mode. I had to keep reminding myself that we were no longer at war, and that there were no kamikazes overhead or roaming enemy submarines underneath the sea.

The cruise across the Pacific to China was uneventful. It seemed both strange and unsettling to go halfway around the world and know that we were not going to some place like Iwo Jima or Okinawa. Seventeen days later, on 23 January 1946, we took on a Chinese pilot to get through the brown, churning Yangtze River, up the river to where it intersected with Whangpoo River. Once successfully navigated, our Chinese friend disembarked and another pilot took his place to move the ship up the winding Whangpoo River. The ships in the river lined up at the buoys in the middle of

221

the river in a chain-like fashion. From the air, it had to look like a snake in the middle of the river. The ships run for miles and were crisscrossed by hundreds of junks and sampans that carry everything from people to merchandise.

From there, we went into Shanghai. On the coast of the East China Sea, north of Formosa and south of Tsingtao was the city of Shanghai, China. We had come to pick up Marines who found themselves stranded there after the war was over. We would again be a transport ship, only this time for the joyful duty of taking Marines home to the States. What could be better than that? For the first time since the USS *NAPA* had been sent to war, she had to pass customs! That was another indicator that we were no longer at war!

I decided to go on shore for my souvenirs. Using my ever faithful pass, I went on shore in Shanghai. I took advantage of the time and did a lot of sightseeing. I also rode rickshaws, my favorite pastime.

Shanghai was an international city and had French, German, Russian and American quarters in its sprawling city, beside the native Chinese. There was a ballroom called the Paramount that catered to the international trade, especially the Americans. Rickshaws were the preferred way of travel to the Paramount. Kapp and I went out that evening to the Paramount Ballroom, and I couldn't believe my ears when I walked into the place. There was an all-Chinese band, Chinese singers, singing in American-accented English to a Glenn Miller piece. If I closed my eyes, I would have sworn that they were really Glenn Miller's Band.

On one side of the wide, lavish ballroom were the girls, all lined up in their pretty dresses and subtle makeup. On the other side were the men. It was the custom for the men to cross the shiny hardwood floor of the ballroom and ask the ladies for a dance, and then escort them back to their side when the dance was over. Kapp and I didn't mind that custom at all! The Chinese ladies were pretty and very good dancers.

Aboard the ship, Chinese merchants were allowed on the forward boat deck, displaying their wares. The USS *NAPA* had its own department store right there on deck! Everyone was buying souvenirs. I bought a roll of silk for my Mom, but I got it in Shanghai itself instead of the impromptu department store.

Another "shopping mall" in Shanghai were the "junks," which were similar to the sampans. These small fishing vessels were the homes for hundreds of Chinese families, who lived, worked, slept, and ate on board. Fish was (and is) a mainstay in China, and the fishermen were an important part of their economical structure. The junk fishermen were tireless, hard workers.

It was, all in all, a quiet trip. It was quite a difference from our previous trips to China. There was little destruction in Shanghai. I've no idea why it was spared, since it was sitting on the coast and the coastal cities were the most vulnerable from the Japanese attacks. But Shanghai, perhaps because it was an international port, showed very little scars from the terrible war that had recently ended.

CHAPTER 28
Home

4 February 1946, we left Shanghai for Tsingtao, China, to pick up Marines, Coast Guard, Seabees, and other Navy personnel to take them home! This, too, was quite a different trip from our previous one when our task was to rescue the Nationalists Chinese and relocate them to relative safety.

It was bitterly cold when we reached Tsingtao. The USS *NAPA* berthed in the same berth it had previously. Two days later, on 6 February, heavy winds and snow storms detained us from departing from Tsingtao. It was so cold we piled on all the clothes we could to ward off the unexpected deep chill. An Oriental winter is unbelievably cold! There was ice in the sea, thin, floating slivers of ice, moving slowly along, bouncing against the ship. I had grown up in Arkansas winters, but even that did not prepare me for a winter day in China. There's nothing that can compare to that, except maybe opening up the refrigerator door and sitting in the freezer overnight.

Finally, on 8 February 1946, we got underway from Tsingtao, China, with troops headed for home! This time, the crowded ship didn't bother me at all. We were taking these troops home, and in the process, we were taking ourselves home!

It took twenty days, but on 24 February 1946, we pulled into the port at San Diego, California. There was a great morale-boosting sign at the harbor when we pulled in. It said: "**WELCOME HOME—**

WELL DONE" and had red-white-blue bunting. We felt we were home! It didn't matter if sunny Southern California was, at that moment, overcast and gray. This was the sunshine! That sign welcoming us home! I watched from the radio shack deck as we disembarked two Navy nurses, 55 Navy officers, 478 enlisted personnel, 43 Marine officers, and 1,029 enlisted Marines. These were the ones we taxied from China to the shores of California.

The next day, we set sail again, but this time not back to the Pacific or China, but to the Panama Canal. We were on our way to Norfolk, Virginia, our final destination. It was quite an experience going from the Pacific Ocean to the Atlantic Ocean in a matter of hours by the way of the Panama Canal. We went slowly through each of the locks until we transversed from one ocean to the other.

The canal was so narrow that, when standing on deck, if I stretched out my hand, I could have touched the sides of the canal. At each lock, the ship would seem to be suspended while the past lock was closed and the oncoming one was opened. We went through them all, and it was an awesome sight to see the large ships looming over the locks, waiting their turns.

There was only one more port-of-call for the mighty USS *NAPA*: Norfolk, Virginia. From the Panama Canal, we went northward in the Atlantic to Norfolk, home of the USS *NAPA*. The ship and its crew had served the country well. Now it was time for rest from the war. The USS *NAPA* was decommissioned, and we were sent home. It seemed anticlimactic after what we had seen and where we had been.

It's been over six decades since that time, yet the memories are as fresh as though it were yesterday. The men who served on Iwo Jima I didn't know individually by name, but they will always be thought of as comrades, friends, and heroes. We walked through fire together, and some of us were left behind, yet we carry those with us always. Freedom isn't free; it is bought with a price, and the price was paid in such places as the Philippines, Guam, Guadalcanal, Okinawa, Saipan, and Iwo Jima. The men who fought there are fewer each year, but it's important not to forget. Wars aren't dry words in a history book, to shake your head over and "armchair quarterback" the battles. Wars are about individuals who are willing to put aside their own safety to ensure the freedom of others. It doesn't matter if the "others" don't believe in war or the cause for

which they fight. The concept of freedom is freedom for all. May
we never forget!

Welcome home! Well Done!

Sign at San Diego, California, 24 February 1946, welcoming the NAPA *home.*

TRAVELS OF THE USS *NAPA*, APA 157

Portland, Oregon
Astoria, Oregon
Seattle, Washington
San Francisco, California
San Pedro Bay, California
Long Beach, California
Port Heuneme, California
Pearl Harbor, Hawaii
Maui, Hawaii
Eniwetok Atoll, Japanese Territory
Saipan, Mariana Islands, Japanese Territory
IWO JIMA, JAPANESE TERRITORY
Guam, Japanese Territory
Ulithi, Caroline Islands, Japanese Territory
Zohnoiiyoru Bank, Japanese Territory
Okinawa, Japanese Territory
Leyte, The Philippines
Manila, The Philippines
Yokohama, Japan
Chefoo, China
Tsingtao, China
Haifong, China
Formosa, China
San Diego, California
Panama Canal, Panama
Norfolk, Virginia

Printed in the United States
138303LV00001B/29/P